Molecular Biology Intelligence Unit

Fibrous Proteins

Thomas Scheibel, PhD
Professor for Biomaterials at the Faculty of Applied Science
Universität Bayreuth, Germany
Bayreuth, Germany

CRC Press
Taylor & Francis Group
Boca Raton London New York

CRC Press is an imprint of the
Taylor & Francis Group, an **informa** business

Fibrous Proteins

Molecular Biology Intelligence Unit

CRC Press
Taylor & Francis Group
6000 Broken Sound Parkway NW, Suite 300
Boca Raton, FL 33487-2742

First issued in hardback 2019

ISBN-13: 978-1-58706-316-9 (hbk)

Visit the Taylor & Francis Web site at
http://www.taylorandfrancis.com

and the CRC Press Web site at
http://www.crcpress.com

Library of Congress Cataloging-in-Publication Data

Fibrous proteins / [edited by] Thomas Scheibel.
p. ; cm. -- (Molecular biology intelligence unit)
Includes bibliographical references and index.
ISBN 978-1-58706-316-9
1. Cytoskeletal proteins. 2. Amyloid. 3. Collagen. 4. Fibrin. 5. Keratin. 6. Tubulins. I. Scheibel, Thomas. II. Series: Molecular biology intelligence unit (Unnumbered)
[DNLM: 1. Scleroproteins. QU 55.3 F443 2008]
QP552.C96F53 2008
572'.67--dc22

2008031585

Dedication

Dedicated to Melanie, Phillip, Rainer and Waltrand

About the Editor...

THOMAS SCHEIBEL is Professor for Biomaterials at the Faculty of Applied Science at the Universität Bayreuth, Germany. Main research interests include fibrous proteins such as spider silks and mussel collagens. He gained several prestigious awards including an "Innovation by Nature" award of the German Ministry of Science and Education in 2007. He is further one of 10 recipients of the 2006 innovation tribute of the Bavarian prime minister. He received his Diploma of biochemistry, Dr. rer. nat. from the Universität Regensburg, Germany and his Habilitation from the Technische Universität München, Germany.

CONTENTS

EDITOR

Thomas Scheibel
Professor for Biomaterials at the Faculty of Applied Science
Universität Bayreuth, Germany
Bayreuth, Germany
Email: thomas.scheibel@uni-bayreuth.de
Chapter 9

CONTRIBUTORS

Note: Email addresses are provided for the corresponding authors of each chapter.

Russell F. Doolittle
Departments of Chemistry & Biochemistry and Molecular Biology
University of California, San Diego
La Jolla, California, USA
Email: rdoolittle@ucsd.edu
Chapter 4

Kenneth H. Downing
Life Sciences Division
Lawrence Berkeley National Laboratory
Berkeley, California, USA
Email: khdowning@lbl.gov
Chapter 7

Matthew J. Harrington
Department of Molecular, Cellular and Developmental Biology (MCDB)
University of California, Santa Barbara
Santa Barbara, California, USA
Email: mjh@umail.ucsb.edu
Chapter 3

David J.S. Hulmes
Institut de Biologie et Chimie des Protéines
Universite de Lyon
Lyon, France
Email: d.hulmes@ibcp.fr
Chapter 2

Keiichi Namba
Graduate School of Frontier Biosciences
Osaka University
Yamadaoka, Suita, Osaka, Japan
Chapter 5

David A.D. Parry
Institute of Fundamental Sciences
Massey University
Palmerston North, New Zealand
Email: d.parry@massey.ac.nz
Chapter 6

Roma N. Rambaran
Department of Chemistry and Biochemistry
School of Life Sciences, John Maynard-Smith Building
University of Sussex
Falmer, Brighton, UK
Chapter 1

Lin Römer
Biomaterials
Universität Bayreuth
Bayreuth, Germany
Chapter 9

František Sehnal
Biology Centre ASCR
České Budějovice, Czech Republic
Email: sehnal@bc.cas.cz
Chapter 8

Louise C. Serpell
Department of Chemistry and Biochemistry
School of Life Sciences, John Maynard-Smith Building
University of Sussex
Falmer, Brighton, UK
Email: l.c.serpell@sussex.ac.uk
Chapter 1

Thomasin A. Smith
Institute of Fundamental Sciences
Massey University
Palmerston North, New Zealand
Chapter 6

Tara Sutherland
CSIRO Entomology
Canberra, Australia
Chapter 8

Ferenc Vonderviszt
Department of Nanotechnology
University of Pannonia
Veszprém, Hungary
and
Institute of Enzymology
Hungarian Academy of Sciences
Budapest, Hungary
Email: von007@almos.vein.hu
Chapter 5

J. Herbert Waite
Department of Molecular, Cellular
and Developmental Biology (MCDB)
University of California, Santa Barbara
Santa Barbara, California, USA
Chapter 3

PREFACE

Fibrous Proteins gives an overview of some of the most important members of the family of fiber-forming proteins concerning structure, function, assembly, and even potential applications as novel biopolymers and biomaterials in technical applications.

Investigating protein fibers has engaged the interest of scientists from different disciplines since the beginning of modern protein science. However, while our knowledge of globular proteins has been increasing over the past decades, fibrillar proteins, although heavily investigated in the 1950s and 1960s, have not re-entered center stage until recently. In recent years, research of fibrillar proteins has become very exciting. New technologies have increased the accessibility and analysis of fibrillar proteins, leading to spectacular insights into assembly mechanisms and material properties, some of which are highlighted herein.

In this book, a number of leading experts in the field describe this highly fascinating topic of protein science. Fibrous proteins are essential building blocks of life, providing scaffolds for our cells, both intracellular and extracellular—in some cases (such as with silks) even outside of organisms. However, despite nearly a century of research into the assembly mechanisms and structures of fibrous protein, we are only recently starting to understand them. In the last years, atomic models for some of the most important fibrous proteins have been proposed. Further, we are increasingly knowledgeable about how the fibrous proteins assemble and how they "function" on a molecular basis.

We would like to present some state-of-the-art information in the area of fibrous proteins science to inspire students as well as advanced researchers and professionals. Future studies will further elucidate the relevance of protein assembly processes in cellular development, cellular structure, and cellular stability. Beyond that, assessing fundamental insights into mechanisms of protein fiber assembly will help to unravel some fundamental principles involved in protein folding diseases (such as Alzheimer's, Parkinson's, Huntington's, Prion's, etc. disease), which will be extremely helpful in disease prevention. Due to the outstanding material properties of some fibrous proteins (especially of those that are employed outside the bodies of organisms), development of products in the area of material sciences can be envisioned, which is of high commercial interest given the fact of rising oil prices and decreasing inventions in "classical" polymer sciences.

Thomas Scheibel, PhD

Acknowledgements

I am grateful to all previous and current members of my lab for "helping hands" and highly inspiring discussions.

Chapter 1

Amyloid Fibrils:
Abnormal Protein Assembly

Roma N. Rambaran and Louise C. Serpell*

Abstract

Amyloid refers to the abnormal fibrous, extracellular, proteinaceous deposits found in organs and tissues. Amyloid is insoluble and is structurally dominated by β-sheet structure. Unlike other fibrous proteins it does not commonly have a structural, supportive or motility role but is associated with the pathology seen in a range of diseases known as the amyloidoses. These diseases include Alzheimer's, the spongiform encephalopathies and type II diabetes, all of which are progressive disorders with associated high morbidity and mortality. Not surprisingly, research into the physicochemical properties of amyloid and its formation is currently intensely pursued. In this chapter we will highlight the key scientific findings and discuss how the stability of amyloid fibrils impacts on bionanotechnology.

Introduction

Amyloid fibrils are formed by normally soluble proteins, which assemble to form insoluble fibers that are resistant to degradation. Their formation can accompany disease and each disease is characterised by a specfic protein or peptide that aggregates. Well known examples of amyloid diseases include Alzheimer's disease, Diabetes type 2 and the spongiform encephalopathies (e.g., Mad cow disease). The amyloid fibrils are deposited extracellularly in the tissues and are thought to have a pathogenic effect[1]. The fibrillar assemblies are inherently stable and structural studies have revealed that they are composed predominantly of β-sheet structure in a characteristic cross-β conformation. Recently, a number of examples of functional amyloid have been identified including a constituent of melanosomes, curli and hydrophobins[2].

The History of Amyloid

The term 'amyloid' was coined initially by Schleiden and then by Virchow in the mid-19th century to describe the iodine stained deposits seen in liver at autopsy. Consequently the deposits were thought to be carbohydrate in nature until their high nitrogen content was later established.[3] Nevertheless, the inaccurately descriptive name was retained for these highly proteinaceous deposits. Further tinctorial properties included the specific binding of amyloid to the dye Congo Red which produced an apple green birefringence when examined between cross polarisers in a light microscope[4] (Fig. 1). This finding suggested that amyloid was fibrillar in structure and subsequent transmission electron micrographs of amyloid confirmed this.[5] Further progress in biochemical and biophysical techniques enabled the isolation of amyloid fibrils from tissues[6] and their now characteristic cross-β structure was interpreted from X-ray fiber diffraction patterns.[7]

*Corresponding Author: Louise C. Serpell—Department of Chemistry and Biochemistry, School of Life Sciences, John Maynard-Smith Building, University of Sussex Falmer BN1 9QG, UK. Email: l.c.serpell@sussex.ac.uk

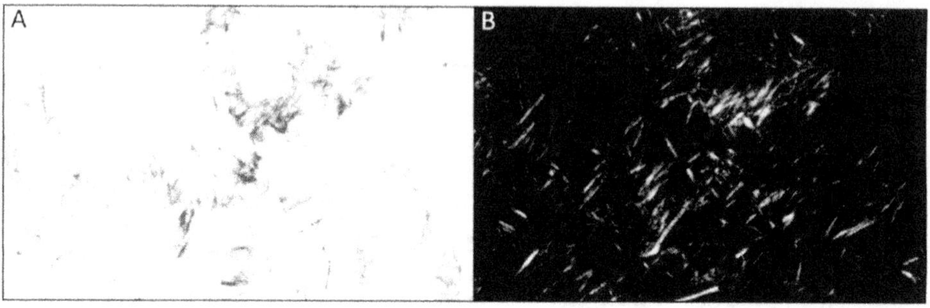

Figure 1. Isolated amyloid fibrils composed of Aα chain fragment of fibrinogen (A) stained with Congo red and visualized by light microscopy and (B) between crossed polars, showing characteristic apple-green birefringence. Figure adapted from reference 83. A color version of this figure is available at www.eurekah.com.

Amino acid composition and sequence analysis of the proteins comprising a range of ex vivo amyloid fibrils revealed that each amyloid disorder was associated with a particular protein or peptide.[8] To date, more than 20 plasma proteins have been identified that form amyloid (Table 1). Interestingly, despite the obvious differences in amino acid sequences and native structure, these amyloidogenic peptides all appear to share a common β-sheet conformation of their polypeptide backbone[7,9] and it is likely that this characteristic confers the fibrillar, proteolytic resistant and insoluble characteristics to all forms of amyloid.

The finding that lysosomal extracts digested immunoglobulin light chain precursor protein into amyloidogenic fragments[10] led to the realisation that many amyloid forming peptides were produced by the proteolytic processing of a precursor protein (Table 1). By 1982, Prusiner put forward the 'prion' hypothesis that described an infectious protein particle capable of causing scrapie (a fatal neurodegenerative disease) in sheep.[11] This protein was amyloidogenic but unlike other amyloid forming proteins it was infective and was termed: prion. Two years later, Aβ, the peptide forming the bulk of amyloid plaques in Alzheimer's diseased brain specimens, was identified and biochemically characterised.[12] In 1986, attention was refocused on prions when a devastating epidemic affected cattle in the UK. These animals were dying of Bovine spongiform encephalopathy (BSE), a prion disease with links to scrapie prion from sheep. Ten years later, a new fatal, neurodegenerative prion disease emerged in humans, variant Creutzfeldt-Jacob disease (vCJD) and the transmissibility of prions from species to species could not be denied as vCJD was strongly linked with exposure to the BSE agent. Interest in prion diseases and amyloid formation (believed to be important to the pathogenesis of the diseases) now escalated.

Earlier biochemical analysis of amyloid deposits had demonstrated that amyloid was not merely composed of fibrils but also contained the serum amyloid protein (SAP) and proteoglycans. Although SAP is believed to have a scavenging function whereby it may stabilize amyloid formation,[13] the role of these proteins in amyloid formation and deposition is yet to be discerned. In 1988, the Pepys laboratory radiolabelled SAP and used it as an in vivo tracer to follow the progress of amyloid disorders in patients.[14]

Three years later the propensity of a peptide to form amyloid was found to be sequence dependent. Polypeptides containing glutamine repeats were found to be associated with some inherited, neurodegenerative amyloidoses.[15] Subsequently, Kelly and coworkers found that purified transthyretin formed fibrils in vitro when its native state was partially unfolded under acidic conditions.[16] This finding suggested that induced conformational changes were sufficient to trigger aggregation into amyloid fibrils and implied that erroneous protein folding or misfolding underpinned the mechanism of amyloid formation, fuelling a new area of intense research in protein chemistry.

Table 1. Some amyloidoses and their respective precursors and amyloidogenic proteins

Disease	Precursor Protein	Amyloid Protein
Alzheimer's disease	Amyloid precursor protein	Aβ peptides
Atrial amyloidosis	Atrial natriuretic factor (ANF)	Amyloid ANF
Spongiform encephalopathies	Prion protein (PrP_c)	PrP_{sc}
Primary systemic amyloidosis	Immunoglobulin light and heavy chains	AL and AH
Senile systemic amyloidosis	Wild-type transthyretin	ATTR
Haemodialysis-related amyloidosis	β_2-microglobulin	$A\beta_2M$
Hereditary nonneuropathic systemic amyloidosis	Lysozyme	ALys
Type II diabetes	Pro-IAPP	IAPP or "amylin"
Injection-localised amyloidosis	Insulin	AIns
Secondary systemic amyloidosis	(Apo) serum amyloid A	Serum amyloid A
Hereditary cerebral amyloid angiopathy	Cystatin C	ACys
Finnish hereditary systemic amyloidosis	Gelsolin	AGel
Familial amyloid polyneuropathy I	Transthyretin variants	ATTR
Familial amyloid polyneuropathy II	Apolipoprotein A1	AApoA1
Ageing pituitary, prolactinomas	Prolactin	APro
Familial amyloidosis	Fibrinogen αA-chain	AFib
British familial dementia	Amyloid Bri Precursor Protein	ABri

Amyloidogenesis: Assembly of Amyloidogenic Proteins

With advances in solid phase protein synthesis techniques in the early nineties it was possible to use synthetic Alzheimer's related peptide, Aβ(1-42), its truncated homologues and other short amyloid forming peptides to produce amyloid-like fibrils.[17] These were physically, morphologically and tinctorially similar to ex vivo amyloid fibrils and enabled the study of amyloid ultrastructure and formation. An extensive range of biophysical tools and techniques including nuclear magnetic resonance (NMR), Circular Dichroism (CD), X-ray fiber diffraction, atomic force and electron microscopy (AFM and EM), Fourier Transform Infrared Spectroscopy (FTIR) have all since contributed to a better understanding of the structure of amyloid fibrils at a molecular level.

Studying the mechanism of formation of amyloid fibrils is of major importance since insights into the mechanisms underlying polymerisation of soluble, monomeric peptide into mature insoluble fibrils may provide researchers with possible therapeutic approaches to halting, reversing or avoiding fibril formation. From milestone studies, it is known that conversion of soluble peptide to insoluble amyloid often involves the production of a partially unfolded intermediate.[18-20] This thermodynamically unfavourable state rapidly progresses to the stable amyloidogenic form and the kinetics of this transition have been studied in vitro using a range of biophysical methods including light scattering, size exclusion chromatography, fluorimetry and ultracentrifugation. The results of such studies propose a nucleation dependent polymerisation model to describe fibril formation, likening the process to crystallization.[21] A heterogenous nucleus ('seed') or peptide micelle form above a critical 'threshold' concentration and fibrils nucleate within these, elongating by irreversibly binding monomers to their free ends.[22,23] Fibril growth may be represented diagrammatically as a lag exponential growth curve where the phase is considerably shortened in the presence of seeds.

In the evolution of mature fibrils, several metastable intermediates have been identified and isolated.[22,24-26] These include very early species: dimers, trimers, tetramers (collectively known as oligomers[27]), Aβ derived diffusible ligands or ADDLs[28] and the later bead-like structures up to 200 nm in length called protofibrils.[29] Alternatively fibril formation may follow an 'offset pathway' without the production of fibrils but instead involving conversion of the intermediates into amorphous deposits.

The dynamics of protein folding/misfolding appear to play a key role in in vitro fibril formation and consequently several molecular mechanisms to explain amyloid formation have been proposed; including the polar zipper[30] and domain swapping models.[31,32] It has been proposed that amyloid formation is a generic property of all peptides[33] since under denaturing conditions many normally globular, nondisease related proteins, have been shown to assemble to form fibrils.[34,35] However, the propensity of a peptide to form amyloid is dependent on several factors including: polypeptide charge, sequence, hydrophobicity and secondary structure. Consequently several algorithms have been developed to predict the propensity and the rate at which different sequences will aggregate and which mutations will result in an increase or decrease of aggregation rate.[36-40] These algorithms will prove useful in determining which regions of a polypeptide chain is specifically involved in fibril formation or the amyloid core structure.

The Structure of the Amyloid Fibril

Amyloid fibrils are insoluble and heterogeneous, so commonly used methods of structure determination are difficult. Therefore, most studies have involved X-ray fiber diffraction, electron microscopy[41] and more recently, solid state nuclear magnetic resonance (ssNMR)[42] and electroparamagnetic resonance.[43-45]

Electron and atomic force microscopy have given insights into the macromolecular structure of amyloid fibrils and have shown that fibrils are long, straight and unbranching (Fig. 2A) and are made up of individual subunits named "protoflaments".[46-48] These may vary in number and are often observed to twist around one another to form the mature fibril.[46,48-50] Synthetic amyloid-like fibrils, may vary in morphology and this may depend upon the assembly conditions.[46,51]

Cryo-electron microscopy and single particle processing of mature amyloid fibrils composed of the SH3 domain of phosphotidylinositol-3′-kinase indicated that a single fibril was comprised of four protofilaments wound around a central core.[2] Consequently, a molecular model of a SH3 amyloid fibril has been generated showing the protofilaments each composed of continuous β-sheet structure.[48] Further cryo-electron microscopy studies revealed that synthetic amyloid fibrils formed by insulin,[50] lysozyme[49] and Aβ(1-42)[52] are also composed of several protofilaments wound around one another. The numbers of protofilaments can differ from 2 to 6.[50]

Early studies on amyloid fibrils revealed that they shared a common X-ray diffraction fingerprint[7] and this was later confirmed in a fiber diffraction study of a number of different ex-vivo and synthetic amyloid fibrils.[9,53,54] The X-ray diffraction pattern given by amyloid fibrils is "cross-β" (Fig. 2B), a diffraction fingerprint first identified for silk from the egg stalk of the lacewing, *Chrysopa.*[55] The pattern indicates that these fibrous molecules share a particular core structure consisting of β-sheet conformation in which the hydrogen bonding direction runs parallel to the fiber axis and the β-strands are perpendicular, much like the rungs of a ladder (Fig. 2). The diffraction pattern consists of two major reflections at 4.7 Å and 10 Å found on orthogonal axes and arising from the hydrogen bonding distances between β-strands and side chain packing between the sheets respectively (see Fig. 2).[9,56] More detailed X-ray diffraction patterns have been obtained from synthetic amyloid fibrils and the additional information obtained has enabled molecular models to be proposed.[57] Generally these models are cross-β in nature and the β-sheet conformation forms the core of the structure. The β-strands are hydrogen bonded along the length of the β-sheet structure and parallel to the fiber axis.[58] The β-sheet ribbons are associated via side chain interactions[57] that serve to stabilise the structure. Very recently, a number of short amyloidogenic

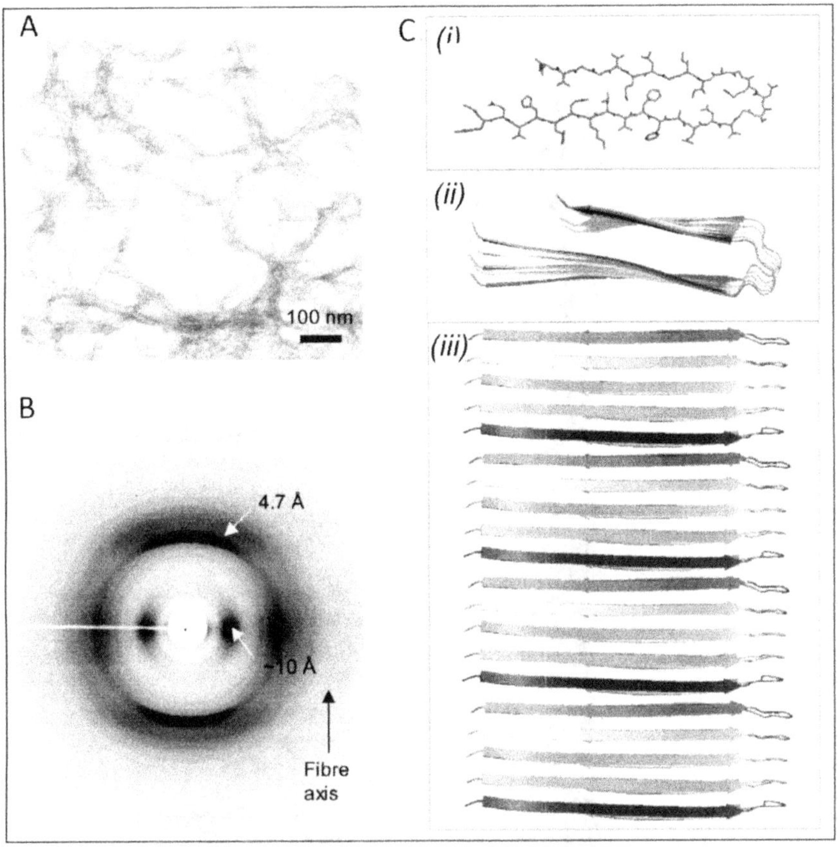

Figure 2. Synthetic amyloid fibrils made from Aβ peptide (A) electron micrograph showing long, straight, unbranching fibrils. B) X-ray fiber diffraction pattern from partially aligned amyloid fibrils showing the characteristic "cross-β" diffraction pattern. C) The structure of the Aβ amyloid fibril interpreted from ssNMR data,[67] showing the top view of the fiber (i and ii) with side chains (i), showing the importance of side chain packing with in the fiber and as a cartoon(ii). The side view (iii) revealing the β-strands running perpendicular to the fiber axis.

peptides have been crystallised this has enabled their atomic structure to be elucidated by X-ray crystallography.[59,60] This work has given insights into the nature of the interactions that drive the association of the β-sheets via the amino acid side chains.

SsNMR structural studies have significantly added to our knowledge of the structure of the amyloid fibril[61] and this work has been complemented by results from EPR on amyloid formed by a number of different disease associated peptides.[45] SsNMR of amyloid fibrils formed by various peptides homologous to regions of the Alzheimer's peptide, Aβ have shown that β-sheet may be arranged parallel or anti-parallel within the protofilaments depending on the properties of the precursor polypeptide.[62-66] Studies of fibrils formed from full-length Aβ, have shown that the peptide folds into a β-bend structure that then associates with other molecules to form parallel, in register β-structure[67,68] (Fig. 2C).

Small angle X-ray scattering has recently contributed insights into both the structure and the assembly of amyloid fibrils[69] revealing for the first time, the oligomeric assembly of insulin fibrils. The SAX study suggested that insulin fibrils are assembled from a helical fibrillation precursor composed of five to six insulin monomers. The mature amyloid fibril is composed of

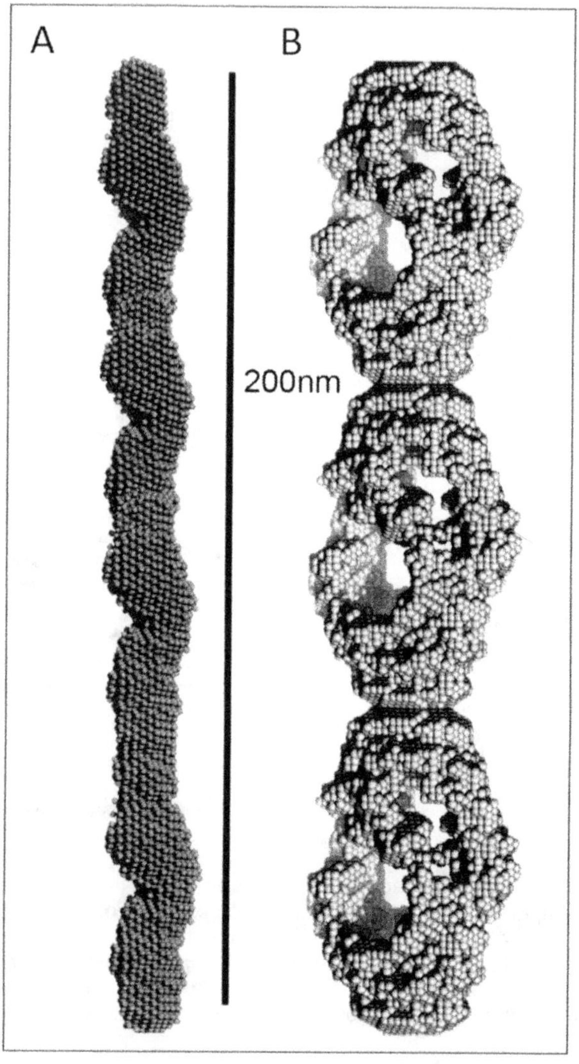

Figure 3. Models of mature protein fibrils based on Small-Angle X-ray scattering solution data. (A) Human alpha-synuclein fibrils and (B) human insulin fibrils.[69] The results suggest that insulin fibrils (B) are formed of three intertwining protofibrils, whereas α-synuclein fibril (A) consist of only one protofibril. Each protofibril is assumed to consist of two intertwining protofilaments. Four and three repeating units are shown for alpha-synuclein and insulin respectively.

three individual filaments that wrap around one another[69] (Fig. 3B). Small angle X-ray scattering from fibrils formed by alpha-synuclein revealed a contrasting structure in which a single filament appears to make up the mature fiber (Vestergaard, personal communication) (Fig. 3A).

Toxicity of Amyloid: The Search for the Toxic Species and Its Mode of Action

It is clear that amyloid deposition is a consequence of the amyloidoses; what is uncertain is whether it is a causative agent in its pathogenesis or a secondary event. The amyloid cascade hypothesis[70] proposed that altered metabolism of amyloid precursor protein (APP) initiates the pathogenesis of Alzheimer's disease (AD) leading to aggregation of Aβ and formation of neuritic plaques. These plaques would cause further pathological changes including the formation of neurofibrillary tangles and compromised synaptic connections ultimately resulting in neuronal cell loss and dementia. However, there was no correlation between the density of plaques and tangles and the severity of AD. Instead, the concentration of soluble Aβ appeared to correlate with cognitive impairment in other studies.[71-73] This finding set the premise for studies suggesting that soluble nonfibrillar intermediates, such as oligomers (20 to >50 kDa globular aggregates, including ADDLs) and protofibrils (curvilinear structures 4-11 nm in diameter and ≤200 nm long) are the actual initiators of AD pathogenesis and that mature fibril formation represents the end point of the disease.

Usually small organic compounds are screened and selected for their binding and inhibitory effects on amyloid precursors or the enzymes involved in their proteolytic processing. These potential anti-amyloid drugs are often highly toxic and have a profound effect on the immune system. Biological ligands such as monoclonal antibodies have been raised against amyloid and are more successful at plaque clearance and reduction in soluble Aβ levels in the CNS.[74] However during phase II trials of an Aβ vaccine, 6% of subjects developed an inflammatory response (meningo-encephalitis) which halted further clinical development. The search therefore for an effective and safe anti-amyloid drug continues. In the interim, several questions remain unanswered: (1) which intermediate species are responsible for the toxicity of amyloid; (2) what is its atomic structure and the exact mechanism of toxicity and (3) how can this toxicity be safely reversed.

"Exploiting Amyloid"

Functional Amyloid

Amyloid fibrils are extremely stable and resistant to degradation. They have been described as having a similar tensile strength to steel,[75] a property that they share with their structural cousin, silk. Recently, a number of nonpathogenic, functional forms of amyloid have been identified in bacteria, fungi, insects and mammals.[2] Curli is a functional fiber found on the surface of bacteria such as *E. coli*. The fibers share structural similarities to amyloid and formed by a protein called CsgA. Fibrillisation is carefully regulated by a protein machine complex.[76,77] Many fungi produce amphipathic proteins called hydrophobins.[78] These have the ability to assemble into β-sheet rich fibrils at air-water interfaces and these are thought to play a protective role in fungal structures such as spores and fruiting bodies. The yeast prions have been suggested to play a functional role, forming cytoplasmic fibrillar assemblies that may be associated with a method of heritable information transfer.[79]

In mammals, a protein called Pmel17 is involved in the biosynthesis of melanin. Recently, the Pmel17 protein has been found to form fibrous structures with all the characteristics of amyloid fibrils and these structures are found at the surface of melanosomes. These are thought to be involved in the templating of melanin biosynthesis.[80]

Bionanotechnological Applications

The amyloid fibril has been increasingly examined for its potential role in forming nanotubular scaffolding for bionanotechnology. The fibrils themselves can be very strong, as previously discussed and these fibers can also be functionalized by assembling fusion proteins[81] or used as a template for binding to metals.[80,82] Baldwin and co-workers[81] assembled a fusion protein composed of a functional cytochrome b_{562} with an amyloidogenic SH3 sequence. The assemblies have the amyloid-like core, displaying functional, folded, globular cytochrome.[81] Nanowires have been fabricated by assembling proteins such as the N-terminal region of the yeast prion, Sup35. Conjugate colloidal gold particles were associated along the fibers using exposed cysteine residues of a variant Sup35,[80] yielding wires around 100 nm in diameter (Fig. 4A). A very short peptide, composed of two phenylalanine residues, assembles to form amyloid-like nanotubes and these may

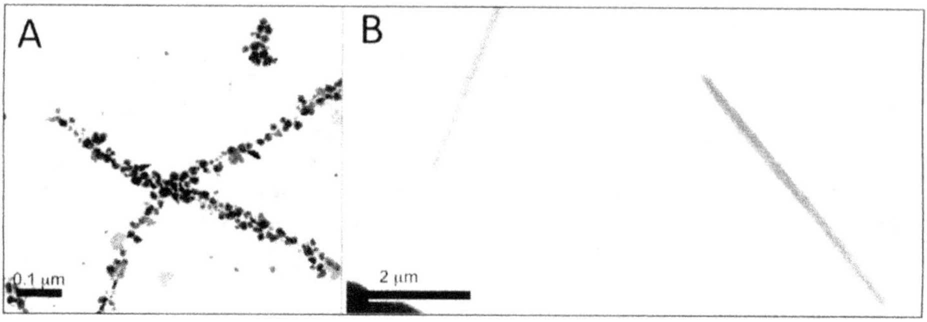

Figure 4. Amyloid-like fibers for bionanotechnology. A) Nanowires based on the N-terminal region of the yeast prion, Sup35. Nanogold was covalently linked to the engineered cysteine residues in the protein and conjugate colloidal gold and silver particles were associated along the fibers to form wires,[80] B) assembly of diphenylalanine to form nanotubes that can be filled with silver to make nanowires.[82]

be functionalised using ionic silver in the centre of the nanotube (Fig. 4B).[82] This work yielded nanowires around 20 nm in diameter.

Conclusions

Amyloid fibrils are extremely strong, highly ordered and organised fibers that can be formed by a large number of proteins and peptides. Their stability and insolubility means that they are useful in a large number of naturally occurring forms as well as bionanotechnology. However, it also means that they are destructive and have the ability to accumulate in the tissues in disease. Although recent key advances have been made in understanding the structure of amyloid-like fibrils, better understanding of the mechanism of assembly is essential for combating amyloid diseases as well as exploiting the potential benefits of protein/peptide design.

Acknowledgements

RNR is supported by a grant from the BBSRC, LCS would like to acknowledge the support of the BBSRC, Alzheimer's research trust, Royal Society and Wellcome Trust. The authors would like to express their thanks to Dr Vestergaard for kind contribution of Figure 3 and to Dr Scheibel for Figure 4A and to Dr Gazit for Figure 4B. They would also like to thank Dr Curtis Rambaran for critical reading of the manuscript.

References

1. Pepys M. Pathogenesis, diagnosis and treatment of systemic amyloidosis. Phil Trans Soc Lond B 2001; 356:203-211.
2. Fowler DM, Koulov AV, Alory-Jost C et al. Functional amyloid formation within mammalian tissue. PLoS Biol 2006; 4(1):e6.
3. Friedrich NaK A. Zur amyloidfrage. Arch Pathol Anat Physiol Klin Med 1859; 16:50-65.
4. Divry P, Florkin M. Sur les proprietes optiques de l'amyloid. Societe de Biologie 1927; 97:180-181.
5. Cohen AS, Calkins E. Electron microscopic observation on a fibrous component in amyloid of diverse origins. Nature 1959; 183:1202-1203.
6. Cohen AS, Calkins E. The isolation of amyloid fibrils and a study of the effect of collagenase and hyaluronidase. J Cell Biol 1964; 21:481-486.
7. Eanes ED, Glenner GG. X-ray diffraction studies on amyloid filaments. J Histochem Cytochem 1968; 16(11):673-677.
8. Glenner G, Ein D, Eanes E et al. Creation of "Amyloid" Fibrils from Bence Jones Proteins. Science 1971; 174:712-714.
9. Sunde M, Serpell L, Bartlam M et al. Common core structure of amyloid fibrils by synchrotron X-ray diffraction. J Mol Biol 1997; 273:729-739.
10. Glenner GG, Terry W, Harada M et al. Amyloid fibril proteins: proof of homology with immunoglobulin light chains by sequence analyses. Science 1971; 172(988):1150-1151.

11. Prusiner SB. Novel proteinaceous infectious particles cause scrapie. Science 1982; 216(4542):136-144.
12. Glenner GG, Wong CW. Alzheimer's disease: initial report of the purification and characterization of a novel cerebrovascular amyloid protein. Biochem Biophys Res Commun 1984; 120(3):885-890.
13. Botto M, Hawkins PN, Bickerstaff MC et al. Amyloid deposition is delayed in mice with targeted deletion of the serum amyloid P component gene. Nat Med 1997; 3(8):855-859.
14. Hawkins PN, Myers MJ, Lavender JP et al. Diagnostic radionuclide imaging of amyloid: biological targeting by circulating human serum amyloid P component. Lancet 1988; 1(8600):1413-1418.
15. La Spada AR, Wilson EM, Lubahn DB et al. Androgen receptor gene mutations in X-linked spinal and bulbar muscular atrophy. Nature 1991; 352(6330):77-79.
16. Colon W, Kelly JW. Partial denaturation of transthyretin is sufficient for amyloid fibril formation in vitro. Biochemistry 1992; 31(36):8654-8660.
17. Serpell L. Alzheimer's amyloid fibrils: structure and assembly. Biochim. Biophys. Acta 2000; 1502:16-30.
18. McParland VJ, Kad NM, Kalverda AP et al. Partially unfolded states of beta(2)-microglobulin and amyloid formation in vitro. Biochemistry 2000; 39:8735-8746.
19. Jahn TR, Parker MJ, Homans SW et al. Amyloid formation under physiological conditions proceeds via a native-like folding intermediate. Nat Struct Mol Biol 2006; 13:195-201.
20. Eakin CM, Berman AJ, Miranker AD. A native to amyloidogenic transition regulated by a backbone trigger. Nat Struct Mol Biol 2006; 13:202-208.
21. Come J, Fraser P, Lansbury PJ. A Kinetic Model for Amyloid Formation In the Prion Diseases, Importance of Seeding. PNAS USA 1993; 90(13):5959-5963.
22. Lomakin A, Chung DS, Benedek GB et al. On the nucleation and growth of amyloid beta-protein fibrils: detection of nuclei and quantitation of rate constants. Proc Natl Acad Sci USA 1996; 93(3):1125-1129.
23. Walsh DM, Hartley DM, Kusumoto Y et al. Amyloid beta-protein fibrillogenesis. Structure and biological activity of protofibrillar intermediates. J Biol Chem 1999; 274(36):25945-25952.
24. Ding TT, Harper JD. Analysis of amyloid-beta assemblies using tapping mode atomic force microscopy under ambient conditions. In: Wetzel R ed. Methods in Enzymology: Amyloid, prions and other protein aggregates. Vol 309: Academic press; 1999:510-525.
25. Roher AE, Baudry J, Chaney MO et al. Oligomerizaiton and fibril asssembly of the amyloid-beta protein. Biochim Biophys Acta 2000; 1502(1):31-43.
26. Harper JD, Wong SS, Lieber CM et al. Assembly of A beta amyloid protofibrils: an in vitro model for a possible early event in Alzheimer's disease. Biochemistry 1999; 38(28):8972-8980.
27. Podlisny MB, Ostaszewski BL, Squazzo SL et al. Aggregation of secreted amyloid beta-protein into sodium dodecyl sulfate-stable oligomers in cell culture. J Biol Chem 1995; 270(16):9564-9570.
28. Lambert MP, Barlow AK, Chromy BA et al. Diffusible, nonfibrillar ligands derived from Abeta1-42 are potent central nervous system neurotoxins. Proc Natl Acad Sci USA 1998; 95(11):6448-6453.
29. Harper JD, Wong SS, Lieber CM et al. Observation of metastable Abeta amyloid protofibrils by atomic force microscopy. Chem Biol 1997; 4(2):119-125.
30. Perutz M, Johnson T, Suzuki M et al. Glutamine Repeats as Polar Zippers: Their Possible Role in Inherited Neurodegenerative Diseases. PNAS USA 1994; 91:5355-5358.
31. Schlunegger MP, Bennett MJ, Eisenberg D. Oligomer formation by 3D domain swapping: a model for protein assembly and misassembly. Adv Protein Chem 1997; 50:61-122.
32. Staniforth RA, Giannini S, Higgins LD et al. Three-dimensional domain swapping in the folded and molten-globule states of cystatins, an amyloid-forming structural superfamily. Embo J 2001; 20(17):4774-4781.
33. Dobson CM. Protein folding and misfolding. Nature 2003; 426:884-890.
34. Malisauskas M, Zamotin V, Jass J et al. Amyloid protofilaments from the calcium-binding protein equine lysozyme: formation of ring and linear structures depends on pH and metal ion concentration. J Mol Biol 2003; 330(4):879-890.
35. Bemporad F, Taddei N, Stefani M et al. Assessing the role of aromatic residues in the amyloid aggregation of human muscle acyphophatase. Protein Science 2006; 15:862-870.
36. De la Paz M, Serrano L. Sequence deteminants of amyloid fibril formation. Proc Natl Acad Sci USA 2004; 101:87-92.
37. DuBay KF, Pawar AP, Chiti F et al. Prediction of the absolute aggregation rates of amyloidogenic polypeptide chains. J Mol Biol 2004; 341:1317-1326.
38. Pawar AP, Dubay KF, Zurdo J et al. Prediction of "aggregation-prone" and "aggregation-susceptible" regions in proteins associated with neurodegenerative disease. J Mol Biol 2005; 350:379-392.
39. Zibaee S, Makin OS, Goedert M et al. A simple algorithm locates beta-strands in the amyloid fibril core of alpha-synuclein, Abeta and tau using the amino acid sequence alone. Protein Sci 2007; 16(5):906-918.

40. Fernandez-Escamilla AM, Rousseau F, Schymkowitz J et al. Prediction of sequence-dependent and mutational effects on the aggregation of peptides and proteins. Nat Biotechnol 2004; 22(10):1302-1306.
41. Makin OS, Sikorski P, Serpell LC. Diffraction to study protein and peptide assemblies. Curr Opin Chem Biol 2006; 10(5):417-422.
42. Tycko R. Insights into the amyloid folding problem from solid-state NMR. Biochemistry 2003; 42:3151-3159.
43. Der-Sarkissian A, Jao CC, Chen J et al. Structural organisation of alpha-synuclein studied by site-directed spin labelling. J Biol Chem 2003; 278:37530-37535.
44. Jayasinghe SA, Langen R. Identifying structural features of fibrillar islet amyloid polypeptide using site-directed spin labeling. J Biol Chem 2004; 279:48420-48425.
45. Torok M, Milton S, Kayed R et al. Structural and dynamic features of Alzheimer's Abeta peptide in amyloid fibrils studied by site-directed spin labelling. J Biol Chem 2002; epub ahead of print.
46. Goldsbury C, Kistler J, Aebi U et al. Watching amyloid fibrils grow by time-lapse atomic force microscopy. J Mol Biol 1999; 285:33-39.
47. Serpell L, Sunde M, Benson M et al. The protofilament substructure of amyloid fibrils. J Mol Biol 2000; 300(5):1033-1039.
48. Jimenez JL, Guijarro JI, Orlova E et al. Cryo-electron microscopy structure of an SH3 amyloid fibril and model of the molecular packing. Embo J 1999; 18(4):815-821.
49. Jimenez JL, Tennent G, Pepys MB et al. Structural diversity of ex vivo amyloid fibrils studied by cryo-electron microscopy. J Mol Biol 2001; 311:241-247.
50. Jimenez JL, Nettleton EJ, Bouchard M et al. The protofilament structure of insulin amyloid fibrils. Proc Natl Acad Sci USA 2002; 99:9196-9201.
51. Goldsbury C, Kistler J, Aebi U et al. Watching amyloid fibrils grow by time-lapse atomic force microscopy. J Mol Biol 1999; 285:33-39.
52. Sachse C, Xu C, Wieligmann K et al. Quaternary structure of a mature amyloid fibril from Alzheimer's Abeta(1-40) peptide. J Mol Biol 2006; 362(2):347-354.
53. Blake C, Serpell L. Synchrotron X-ray studies suggest that the core of the transthyretin amyloid fibril is a continuous b-helix. Structure 1996; 4:989-998.
54. Inouye H, Fraser P, Kirschner D. Structure of β-Crystallite Assemblies by Alzheimer β Amyloid Protein Analogues: Analysis by X-Ray Diffraction. Biophys J 1993; 64(feb):502-519.
55. Geddes AJ, Parker KD, Atkins EDT et al. "Cross β" Conformation in Protein. J Mol Biol 1968; 32:343-358.
56. Makin OS, Sikorski P, Serpell LC. Diffraction to study protein and peptide assemblies. Curr Opin Chem Biol 2006; 10(5):417-422.
57. Makin OS, Atkins E, Sikorski P et al. Molecular basis for amyloid fibril formation and stability. Proc Natl Acad Sci USA 2005; 102:315-320.
58. Makin OS, Serpell LC. Structures for amyloid fibrils. FEBS J 2005; 272:5950-5961.
59. Nelson R, Eisenberg D. Recent atomic models of amyloid fibril structure. Curr Opin Struct Biol 2006; 16(2):260-5.
60. Sawaya MR, Sambashivan S, Nelson R et al. Atomic structures of amyloid cross-beta spines reveal varied steric zippers. Nature 2007; 447(7143): 453-7.
61. Tycko R. Progress towards a molecular-level structural understanding of amyloid fibrils. Curr Opin Struct Biol 2004; 14:96-103.
62. Naito A, Kamihira M, Inoue R et al. Structural diversity of amyloid fibril formed in human calcitonin as revealed by site-directed 13C solid-state NMR spectroscopy. Magn Reson Chem 2004; 42(2):247-257.
63. Gordon DJ, Balbach JJ, Tycko R et al. Increasing the amphiphilicity of an amyloidogenic peptide changes the beta-sheet structure in the fibrils from antiparallel to parallel. Biophys J 2004; 86:428-434.
64. Balbach JJ, Petkova AT, Oyler NA et al. Supramolecular structure in full-length Alzheimer's b-amyloid fibrils: evidence for parallel b-sheet organisation from solid state nuclear magnetic resonance. Biophys J 2002; 83:1205-1216.
65. Petkova AT, Buntkowsky G, Dyda F et al. Solid state NMR reveals a pH-dependent antiparallel beta-sheet registry in fibrils formed by a beta-amyloid peptide. J Mol Biol 2004; 335:247-260.
66. Petkova AT, Leapman RD, Guo Z et al. Self-propagating, molecular-level polymorphism in Alzheimer's beta-amyloid fibrils. Science 2005; 307:262-265.
67. Petkova AT, Ishii Y, Balbach JJ et al. A structural model for Alzheimer's beta-amyloid fibrils based on experimental constraints from solid state NMR. Proc Natl Acad Sci USA 2002; 99:16742-16747.
68. Luhrs T, Ritter C, Adrian M et al 3D structure of Alzheimer's amyloid-beta(1-42) fibrils. Proc Natl Acad Sci USA 2005; 102(48):17342-17347.
69. Vestergaard B, Groenning M, Roessle M et al. A helical structural nucleus is the primary elongating unit of insulin amyloid fibrils. PLoS Biol 2007; 5(5):e134.

70. Hardy J, Allsop D. Amyloid deposition as the central event in the aetiology of Alzheimer's disease. Trends Pharmacol Sci 1991; 12(10):383-388.
71. Lue LF, Kuo YM, Roher AE et al. Soluble amyloid beta peptide concentration as a predictor of synaptic change in Alzheimer's disease. Am J Pathol 1999; 155(3):853-862.
72. Naslund J, Haroutunian V, Mohs R et al. Correlation between elevated levels of amyloid beta-peptide in the brain and cognitive decline. Jama 2000; 283(12):1571-1577.
73. Wang J, Dickson DW, Trojanowski JQ et al. The levels of soluble versus insoluble brain Abeta distinguish Alzheimer's disease from normal and pathologic aging. Exp Neurol 1999; 158(2):328-337.
74. Hock C, Konietzko U, Papassotiropoulos A et al. Generation of antibodies specific for beta-amyloid by vaccination of patients with Alzheimer's disease. Nature Medicine 2002; 8:1270-1275.
75. Smith JF, Knowles TP, Dobson CM et al. Characterization of the nanoscale properties of individual amyloid fibrils. Proc Natl Acad Sci USA 2006; 103(43):15806-15811.
76. Chapman MR, Robinson LS, Pinkner JS et al. Role of Escherichia coli curli operons in directing amyloid fiber formation. Science 2002; 295(5556):851-855.
77. Robinson LS, Ashman EM, Hultgren SJ et al. Secretion of curli fiber subunits is mediated by the outer membrane-localized CsgG protein. Mol Microbiol 2006; 59(3):870-881.
78. Sunde M, Kwan AH, Templeton MD et al. Structural analysis of hydrophobins. Micron 2007; epub ahead of print.
79. Uptain SM, Lindquist S. Prions as protein-based genetic elements. Annu Rev Microbiol 2002; 56:703-741.
80. Scheibel T, Parthasarathy R, Sawicki G et al. Conducting nanowires built by controlled self-assembly of amyloid fibers and selective metal deposition. Proc Natl Acad Sci USA 2003; 100:4527-4532.
81. Baldwin AJ, Bader R, Christodoulou J et al. Cytochrome display on amyloid fibrils. Journal of the American Chemical Society 2006; 128(7):2162-3.
82. Reches M, Gazit E. Casting metal nanowires within discrete self-assembled peptide nanotubes. Science 2003; 300:625-627.
83. Serpell LC, Benson M, Liepnieks JJ et al. Structural analyses of fibrinogen amyloid fibrils. Amyloid 2007; 14(3):199-203.

Chapter 2

Vertebrate Collagens—Structures, Functions and Biomedical Applications

David J.S. Hulmes*

Abstract

The structures and functions of members of the vertebrate collagen superfamily are presented, including the fibrillar and fibril-associated collagens, beaded filament collagen, basement membrane and associated collagens, short chain collagens, transmembrane collagens and related collagen-like proteins. Recent progress in the determination of the three-dimensional structure of collagen fibrils is described, as are the various mechanisms that control collagen biosynthesis and assembly both in vitro and in vivo. Finally, the many biomedical applications of collagens are reviewed, including the use of natural, recombinant and biomimetic molecules, in different tissue-specific contexts.

Introduction

Collagens and collagen-like molecules together make up a protein superfamily which, in humans, currently constitutes about 50 members.[1-3] Since by definition each molecule is a trimer, where all three polypeptide chains are not necessarily identical, the corresponding number of genes is much greater. To be a member of the collagen superfamily, each chain must contain one or more regions characterized by the repeating amino acid motif (Gly-X-Y), where X and Y can be any amino acid. This motif allows the chains to form a right handed triple-helical structure (Fig. 1), with all glycine residues buried within the core of the protein and residues X and Y exposed on the surface. Depending on the genetic type of collagen, the triple-helical motif can be a major or minor part of the molecule, other regions consisting of different noncollagenous domains. Whether a particular member of the superfamily is called a collagen or not, as opposed to a collagen-like protein, is largely dependent on how it was discovered, collagens (types I to XVIII) traditionally being assigned to structural roles and collagen-like proteins to other biological functions. But the distinction is becoming blurred, with new functions being discovered for both collagenous and noncollagenous domains and also some collagens being known by alternative names. Members of the superfamily can be grouped into the following categories: fibrillar collagens; fibril associated and related collagens; beaded filament collagen VI; basement membrane and associated collagens; short chain collagens VIII and X and related proteins; transmembrane collagens and collagen-like proteins; collectins and ficolins; other collagens and collagen-like proteins (Table 1; Fig. 2).

*David J.S. Hulmes—Institut de Biologie et Chimie des Protéines, CNRS/Université de Lyon UMR 5086, IFR 128 Biosciences Gerland-Lyon Sud, 7 passage du Vercors, 69367 Lyon cedex 7, France. Email: d.hulmes@ibcp.fr

Fibrous Proteins, edited by Thomas Scheibel. ©2008 Landes Bioscience.

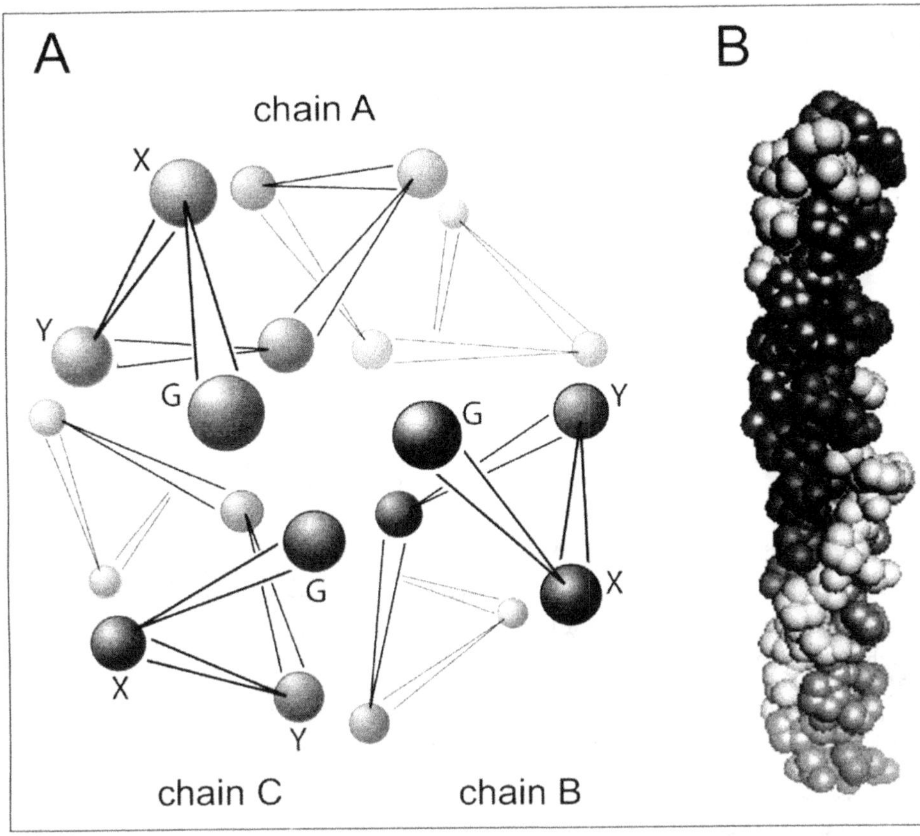

Figure 1. Triple-helical structure of collagen. A) Viewed along the molecular axis (α-carbons only) showing the paths of the individual polypeptide chains and the locations of residues in the Gly-X-Y triplets (G = Gly; from ref. 129 with permission). B) Viewed from the side (molecule tilted towards the reader at the top; space-filling representation) showing the right handed helical twist.

Structures and Functions

Fibrillar Collagens

Fibrillar collagens[4,5] give rise to the classical collagen fibrils seen by electron microscopy, characterized by a repeating banding pattern with a so-called D periodicity of 64 to 67 nm, depending on the tissue. The classical fibrillar collagens include: type I, the most widely occurring collagen found in skin, tendon, bone, cornea, lung and the vasculature; type II, which has a more specific tissue distribution being limited essentially to cartilage; type III, found in relatively elastic tissues such as embryonic skin, lung and blood vessels; type V, found as a quantitatively minor collagen in association with collagen I, with particularly high amounts in cornea; and type XI, a quantitatively minor component of cartilage in association with collagen II.

As in all collagens, each fibrillar collagen molecule consists of three polypeptide chains, called α chains. Molecules can be homotrimeric, consisting of three identical a chains, as in collagens II and III, or heterotypic, consisting of up to three genetically distinct α chains. Individual α chains are identified by the following nomenclature: αn(N), where N is the Roman numeral indicating collagen type and n is the number of the α chain. Thus the chain composition of collagen II is $[\alpha1(II)]_3$, while that of collagen I, a heterotrimer with two identical α1 chains and a third

Table 1. Collagens and collagen-like proteins in vertebrates

Sub-Family	Members
Fibrillar collagens	Types I, II, III, V, XI, XXIV and XXVII
Fibril associated and related collagens	Types IX, XII, XIV, XVI, XIX, XX, XXI and XXII
Beaded filament forming collagen	Type VI
Basement membrane and associated collagens	Type IV, VII, XV and XVIII
Short chain collagens and related proteins	Types VIII and X; C1q; hibernation-related proteins HP-20, HP-25 and HP-27; emilins 1 and 2; adiponectin; CTRPs 1-7: inner ear (saccular) collagen
Transmembrane collagens and collagen-like proteins	Types XIII, XVII, XXIII and XXV/CLAC-P; ectodysplasins; macrophage scavenger receptors I-III; MARCO; SRCL; gliomedin; CL-P1
Collectins and ficolins	Mannan binding protein; surfactant proteins A and D; conglutinin; CL-43; CL-46; CL-L1; CL-P1; L-, M- and H-ficolins
Other collagens and collagen-like proteins	Emu1; collagen XXVI/Emu2; collagen XXVIII; acetylcholinesterase tail subunit

distinct α2 chain, is $[\alpha1(I)]_2\alpha2(I)$. While collagen I has only two different α chains, collagen V has three, leading to different chain stoichiometries such as $[\alpha1(V)]_2\alpha2(V)$ and $\alpha1(V)\alpha2(V)\alpha3(V)$. Collagen XI has three different α chains, where the α3 chain is essentially the same as α1(II). In addition, there are several variants of the α1(XI) and α2(XI) chains due to alternative splicing, corresponding to the variable region, of the corresponding mRNAs. To even further increase diversity of these quantitatively minor fibrillar collagens, hybrid forms of collagen V and XI molecules have also been reported.

The common characteristic of the classical fibrillar collagens is a long central triple-helical region in each α chain, consisting of a continuous $(Gly\text{-}X\text{-}Y)_n$ repeat, where n is 337 to 343 (depending on collagen type). In the case of collagens I, II and III, this region is flanked by short nonhelical regions called telopeptides, typically about 20 residues in length, at both N- and C-termini. All fibrillar collagens are synthesized in the form of soluble precursor molecules, called procollagens, with large N- and C-terminal propeptide domains. The C-propeptides are removed during the later stages of biosynthesis, usually by specific metalloproteinases, leaving the short C-telopeptides. The extent of N-terminal processing depends on collagen type. While for procollagens I, II and III the N-propeptides are completely removed (albeit relatively slowly in the case of procollagen III) leaving short N-telopeptides, N-terminal processing of procollagens V and XI leaves large N-terminal extensions which help modulate fibril formation.

Little is known about the recently discovered fibrillar-like collagens XXIV and XXVII, though these appear to be associated with types I and II containing tissues respectively. They differ from the classical fibrillar collagens in that the (Gly-X-Y) region is relatively short (329 triplets) and this region is interrupted by one (collagen XXIV) or two (collagen XVII) short imperfections in the (Gly-X-Y) repeat. Collagen XXVII has recently been shown to form nonstriated filamentous structures.[6]

Mutations in fibrillar collagen genes are associated with several heritable disorders, including osteogenesis imperfecta (collagen I), Ehlers Danlos syndrome type VII (collagen I), Ehlers-Danlos syndrome type IV (collagen III), classical Ehlers-Danlos syndrome (collagen V), spondyloepiphyseal dysplasia (collagen II), hypochondrogenesis (collagen II), achondrogenesis type II (collagen II), Kniest dysplasia (collagen II), Stickler syndrome (collagens II, XI), Marshall syndrome (collagen XI) and otospondylomegaepiphyseal dysplasia (collagen XI) (see www.le.ac.uk/genetics/collagen and www.hgmd.cf.ac.uk/ac).

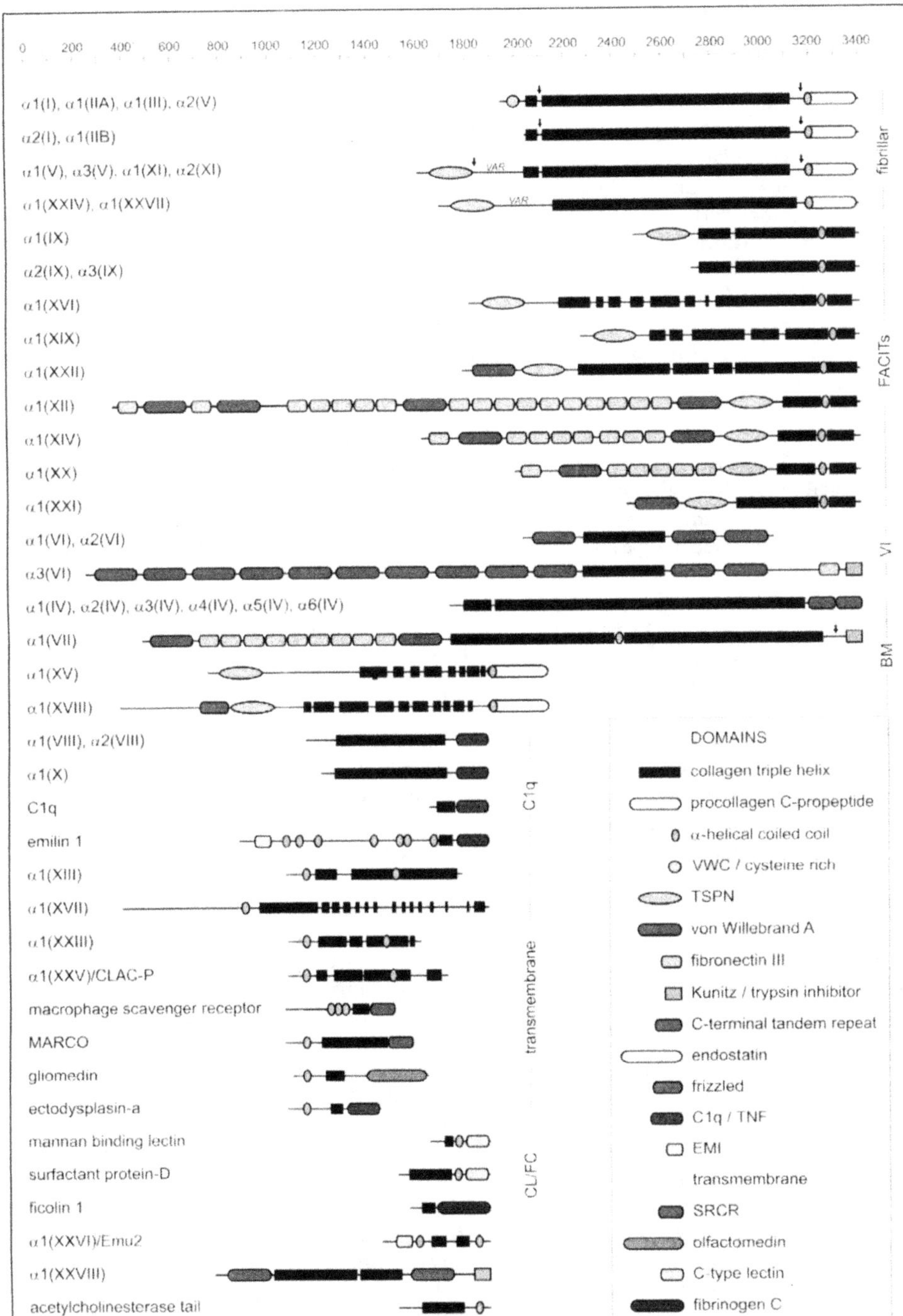

Figure 2. Domain structure of vertebrate collagens and representative collagen-like proteins. Minor interruptions in the triple-helical regions are not indicated. Procollagen forms are shown for the fibrillar collagens and collagen VII, with enzymatic processing sites indicated by arrows. A color version of this image is available at www.eurekah.com.

Nonfibrillar Collagens

The collagens known as FACITs (Fibril Associated Collagens with Interrupted Triple helices; types IX, XII, XIV, XVI, XIX, XX, XXI, XXII) have become a relatively large group within the collagen superfamily.[2] Strongest homology within the group is in the C-terminal collagenous (COL1) and noncollagenous (NC1) domains and all molecules have a thrombospondin N-terminal-like domain (also found in collagens V and XI) just before the first collagenous domain (starting from the N-terminus). The founder FACIT, collagen IX, is an important component of cartilage collagen fibrils, along with collagens II and XI. Unlike the latter however, collagen IX does not itself self-assemble into fibrils. The collagen IX molecule is composed of 3 relatively short COL domains (which contain interruptions in the (Gly-X-Y)n repeat) and 4 NC domains. The molecule is an α1(IX)α2(IX)α3(IX) heterotrimer and the α2(IX) chain sometimes carries a covalently linked sulfated glycosaminoglycan (GAG) chain, thereby making collagen IX a part-time proteoglycan. Collagen IX molecules are covalently linked to the surface of collagen II/XI cartilage fibrils, with the large N-terminal NC4 domain of the α1(IX) chain (which is itself subject to tissue-specific regulation due to the use of alternative transcriptional start sites) projecting into the rest of the ECM.[7] Collagen IX resembles collagens XVI, XIX and XXII in containing three or more collagenous domains, though there is no evidence that these other FACITs associate specifically with collagen fibrils. Collagen XII, an $[\alpha1(XII)]_3$ homotrimer, contains just two collagenous domains, which together account for only about 10% of the entire molecule, the rest being made up of multiple fibronectin type III and von Willebrand factor A domains. Different splice-variants exist, one of which carries a covalently-linked GAG chain, like collagen IX. Collagen XII is also specifically associated with the collagen fibril surface and resembles collagens XXIV, XX and XXI, the main differences being in the size of the noncollagenous regions. In addition to linking collagen fibrils to the rest of the extracellular matrix, possible functions of FACITs include facilitating tissue remodeling,[8] stabilizing tissue junctions[9] and controlling cell quiescence and differentiation[10] and hearing.[11]

Collagen VI is a relatively ubiquitous collagen with important roles in maintaining tissue integrity.[12,13] The molecules contain relatively short triple-helical regions, about one third the length of fibrillar collagens, with large N- and C-terminal regions made up mostly of von Willebrand factor A domains. Each molecule is a heterotrimer α1(VI)α2(VI)α3(VI). In the case of the α3(VI) chain, the triple-helical region makes up less than 10% of the entire amino acid sequence. As with several other collagens (e.g., types II, IX, XI, XII, XIII) further diversity in the α chains is introduced at the transcriptional level through the formation of alternatively spliced variants. Collagen VI molecules assemble into so-called beaded filaments, with a periodicity of 110 nm. Unlike fibrillar collagens, there is no enzymatic processing of the N- and C-terminal noncollagenous regions. In addition, unlike all other collagens, supramolecular assembly begins inside the cell with the formation of anti-parallel dimers and tetramers. Tetramers stabilized by disulphide cross-linking then associate in the extracellular matrix to form beaded filaments or microfibrils.

The nonfibrillar collagen about which most is known is type IV or basement membrane collagen.[14] Basement membranes (or basal laminae) are specialized structures found at tissue boundaries, underlying epithelial, endothelial, fat, muscle and nerve cells. Seen by electron microscopy, they are relatively thin sheets (typically 40-50 nm) composed of collagen IV, laminins, heparin sulfate proteoglycans and nidogens. Collagen IV molecules are longer than the fibrillar collagens and they contain several discontinuities in the (Gly-X-Y) repeat. Also, unlike fibrillar collagens, there is no precursor form of collagen IV and the N- and C-terminal extensions, called respectively the 7S and NC1 domains, are intimately involved in supramolecular assembly. Collagen IV molecules associate via tetramerization of 7S domains, dimerization of NC1 domains, triple-helical interactions and interactions with other basement membrane components to form an open meshwork structure. This structure is stabilized by disulphide bonding in the 7S region, as well as by lysine-derived crosslinks initiated by lysyl oxidase. There are six different collagen IV α chains, resulting in three types of heterotrimer: $[\alpha1(IV)]_2\alpha2(IV)$, α3(IV)α4(IV)α5(IV) and $[\alpha5(IV)]_2\alpha6(IV)$. Mutations in collagen IV or autoimmune recognition of collagen IV are associated with the kidney disorders Alport's and Goodpasture's syndrome, respectively.[14]

Collagen VII has the longest triple-helical region amongst vertebrate collagens, about 420 nm in length, including interruptions and is flanked by noncollagenous NC1 and NC2 domains at the N- and C-terminus, respectively. This collagen is found underlying the basement membrane at the dermal-epidermal junction, in the form of anchoring filaments. Mutations in collagen VII are the cause of severe blistering disorders called epidermolysis bullosa.[15] Within anchoring filaments, about two molecules in length, bundles of collagen VII molecules aligned in register are arranged tail to tail with a short C-terminal overlap. Anti-parallel assembly is triggered by proteolytic processing of the NC2 domain, while the large NC1 domain remains intact and interacts with collagen IV and laminin V.[16]

Collagens XV and XVIII consist of several collagenous domains and together are referred to as multiplexins (for multiple triple helices with interruptions). Both are associated with basement membranes and both carry covalently linked glycosaminoglycan chains and are therefore also proteoglycans. In addition, both include a C-terminal fragment, endostatin, which is proteolytically cleaved from the molecule and has anti-angiogenic properties.[2]

Collagens VIII and X, the so-called short chain collagens, have specific tissue locations, underlying endothelial cells and in the hypertrophic zone of cartilage during endochondral ossification, respectively. The triple-helical region, which contains many interruptions, is about half the length of a fibrillar collagen. While collagen X is an $\alpha1(X)$ homotrimer, collagen VII, with both $\alpha1(VIII)$ and $\alpha2(VIII)$ chains, appears to occur as both homotrimers and heterotrimers. Molecules have both N- and C-terminal noncollagenous regions, the latter corresponding to the widely occurring C1q domain.[17] There is no proteolytic processing of these regions, rather they help in the formation of hexagonal supramolecular networks in, for example, Descemet's membrane of the cornea (collagen VIII[18]) and calcifying cartilage (collagen X[19]).

The transmembrane collagens (types XIII, XVII, XXIII, XXV/CLAC-P) have diverse functions in cell-cell and cell-matrix interactions.[20] Collagen-like membrane proteins include ectodysplasins (involved in the development of hair follicles, teeth and sweat glands[21]), macrophage scavenger receptors and related proteins[22] and gliomedin (involved in nerve development[23]). The collagen-like proteins of the immune system (C1q, collectins, ficolins) have collagen-like stalks and globular heads, the former leading to self-association into bundles containing up to six molecules.[24] In addition to collagens VIII and X, collagen-like proteins containing a C1q-like domain include hibernation-related proteins,[25] adiponectin (implicated in diabetes, obesity, cancer and cardiovascular disease[26]), C1qTNF-related proteins (CTRPs ; involved in inflammation and immunity[27]), inner ear (saccular) collagen[28] and emilins-1 and -2 (involved in elastic tissues[29]). Emilins are part of a larger group of proteins containing the so-called EMI domain, which also includes Emu1 and Emu2 (also called collagen XXVI).[29] Lastly, collagen XVIII is associated with nerve tissue[30] while the acetylcholinesterase complex has a collagen-like tail.[31]

Collagen Fibril Structure

The three-dimensional structure of collagen fibrils has been a subject of debate for over four decades, but a clear picture is now emerging. While earlier studies were carried out on the relatively large diameter fibrils (up to 500 nm) found in tendons, more recent studies have revealed insights into the structures of narrower diameter fibrils (down to 20 nm) found in other connective tissues. A number of points are clear. First, there is universal agreement that molecules within fibrils are staggered by multiples of the D repeat (Fig. 3). It is convenient to think of a fibrillar collagen molecule as made up of five segments (numbered 1 to 5) where segments 1 to 4 are of length D and segment 5 is about half this length (the molecular length being about 4.5 D). Within any one D period of a fibril, there are equal numbers of segments 1 to 5. Different models for the three-dimensional structures of fibrils differ in how these segments are arranged in a fibril cross-section. Second, most fibrils are heterotypic in nature, consisting of different genetic types of collagen (see below). This accounts for the observation that the precise value of the D period depends on the tissue, being

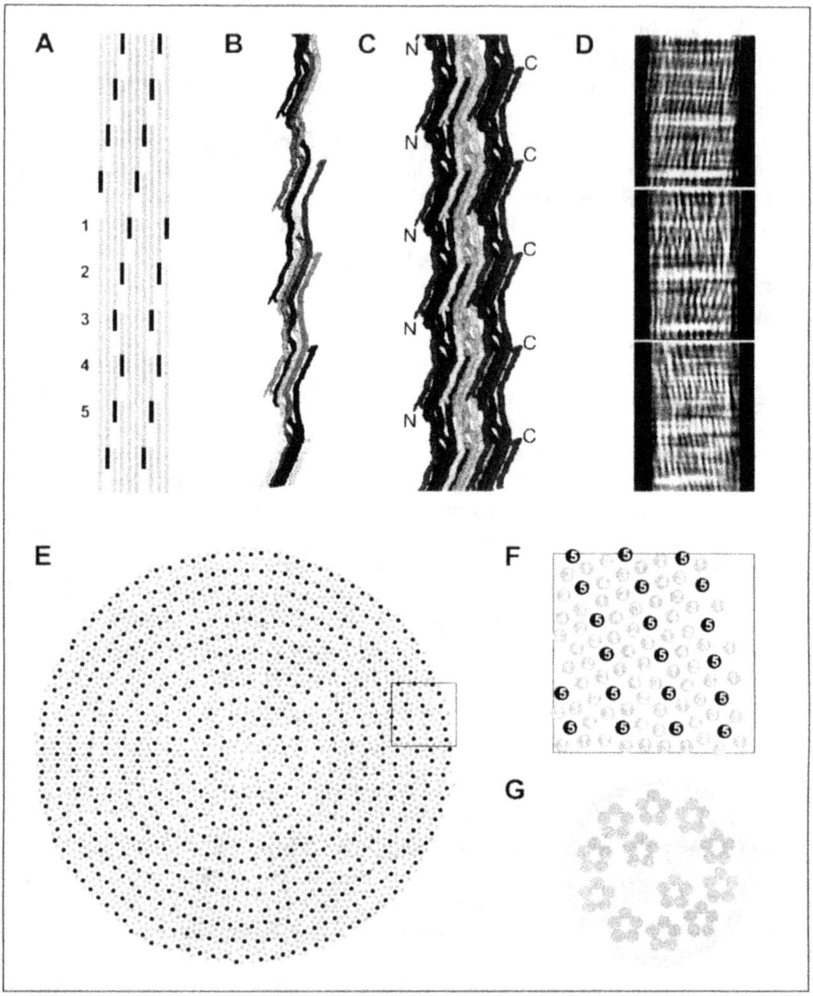

Figure 3. Collagen fibril structure. A-D) longitudinal views. E-G) transverse views. A) D-staggered arrangement of collagen molecules. Each molecule is represented as four segments of length D and a fifth segment (shown in black) of length 0.5 D. B) Microfibrillar arrangement of molecules within large diameter fibrils, from reference 35 with permission. Molecular segments are shown as follows: 1, grey; 2, red; 3, green; 4, blue; 5, yellow. C) Interdigitation of neighboring microfibrils within the fibril, from reference 35 with permission. Each microfibril is indicated by a uniform color. D) Virtual slices through a 3D reconstruction of a narrow diameter corneal fibril showing the helicoidal arrangement of microfibrils, from reference 34 with permission. Top, upper surface; middle, central slice; bottom, lower surface. E) Cross-section of a large diameter fibril (thickness D) showing individual molecules arranged as radially oriented crystalline regions, from reference 36 with permission. Segments 5 are shown in black. F) enlarged view of the boxed region in (E) showing molecular segments corresponding to the structure determined by reference 35; G) cross-section of a small diameter collagen II/XI fibril from cartilage showing individual microfibrils (i.e., groups of 5 molecules) with a 10 + 4 organization. Structures shown in green represent putative collagen XI microfibrils with the large N-terminal domain coating the fibril surface. Based on reference 38. A color version of this figure is available at www.eurekah.com.

about 64 nm in skin and 67 nm in tendon for example, interactions between different collagen types leading to different degrees of tilting of the molecules with respect to the fibril axis.[32] This tilt has been visualized directly by electron microscopy,[33] electron tomography[34] and indirectly by X-ray diffraction[35] and varies from about 15° in skin fibrils to 4° in tendon fibrils (Fig. 3). Third, normal fibrils are approximately circular in cross-section (though this is not the case in certain pathological situations, see below) and this combined with the tilt leads to fibrils having an overall helicoidal organization.

The lateral packing of molecules within collagen fibrils (i.e., perpendicular to the fibril axis) shows features of both order and disorder. This is shown most vividly in X-ray diffraction patterns of collagen fibrils from rat tail tendon, where both diffuse and sharp peaks are seen. As the distance of the peaks from the origin is inversely related to actual spacings within the fibril, it is usual to refer to peak positions as the reciprocal of the corresponding spacing. Thus the diffuse peak at about 1/1.3 nm^{-1} corresponds to short range (liquid-like) order in the intermolecular spacing, while the sharp peak at 1/3.8 nm corresponds to long range order in the arrangement of groups of molecules, usually referred to as microfibrils.[36] The existence of a five-stranded microfibril as the basic building block of collagen fibrils was first suggested by Smith.[37] Recent three-dimensional structure determination of the crystalline regions in large diameter collagen fibrils by X-ray fiber diffraction has shown the presence of microfibrillar-like features,[35] though the details of the structure are more complex than originally thought. First, the molecular tilt is confined to the overlap region. Second, instead of continuous five-stranded microfibrils running throughout the fibril length, molecules in neighboring microfibrils interdigitate and form covalent cross-links (Fig. 3) thereby stabilizing the structure of the entire fibril. The average lateral extent of the crystalline regions rat tail tendon collagen fibrils is about 43 nm, compared to fibril diameters of several hundred nm. This is consistent with the presence of both ordered and disordered regions in the fibril cross-section, with the crystalline regions being oriented radially with respect to the fibril axis.[36] The helicoidal structure has been directly visualized by electron tomography of corneal collagen fibrils,[34] which have uniform narrow diameters of about 36 nm, as shown by "slices" through the 3D reconstructions on opposite sides of the fibril (Fig 3). In addition, both of these studies[34,35] showed evidence for noncollagenous molecules (such as proteoglycans) bound to specific sites in the structure. More recently, electron tomography has also been used to investigate the structure of thin, heterotypic collagen fibrils (diameter 16 nm) isolated from cartilage,[38] where rotational averaging indicated the presence of a microfibrillar substructure with 4 microfibrils in the core surrounded by an outer layer of 10 microfibrils (Fig 3). The common theme emerging from all of these studies is a structure made up of concentric layers of microfibrils, in the form of concentric tubes. Such an organisation is consistent with the tendency of fibrils to form kinks when bent.[39]

Biosynthesis and Assembly of Fibrillar Collagens

Collagen biosynthesis has been studied in greatest detail with regard to the fibrillar collagens. The process is complex, involving numerous intracellular and extracellular steps, all of which contribute to the structure and biomechanical properties of the final fibrils. The first event following synthesis of procollagen α chains on the ribosome is their import into the rough endoplasmic reticulum. There they undergo a series of posttranslational modifications resulting in the assembly of procollagen molecules.[40,41] These steps include: modification of proline residues to hydroxyprolines,[42,43] modification of lysines to hydroxylysines,[44] N- and O-linked glycosylation,[45] trimerisation, disulphide bonding, prolyl cis-trans isomerization and folding of the triple-helix.[46-48] Molecules then transit the Golgi network where they are packaged into secretory vesicles prior to export into the extracellular matrix.[49] Procollagen processing (removal of N- and C-propeptides by specific enzymes including members of the ADAMTS,[50] tolloid[51] and furin[52] families) occurs during the later stages of intracellular transport or shortly after secretion, followed by spontaneous assembly of fibrils. Finally, fibrils are stabilized by the formation of covalent cross-links (mediated by lysyl oxidases,[53] tissue transglutaminase[54] or non-enzymatic glycation[55]).

The formation of collagen fibrils in vitro has also been a subject of extensive research (see 56). Initial observations were on the reconstitution of fibrils from cold solutions of collagen in dilute acid. The major fibrillar collagens are soluble at low pH, for example in dilute acetic acid. When the pH is adjusted to around neutral and the temperature raised to around physiological, fibril formation occurs spontaneously resulting in banded fibrils. The structure of the resulting fibrils in influenced by several parameters, including buffer composition and temperature,[57] order of the initiating procedure (neutralizing the pH before or after raising the temperature;[58]), intactness of the N- and C-telopeptides, the presence of other types of collagen and the presence of macromolecules other than collagens (see below).

The telopeptides play a crucial role.[56] Commercially available fibrillar collagens are of two types, either acid soluble or pepsin soluble. In acid soluble collagen, the telopeptides are usually intact and these help both to initiate fibril formation and to produce long cylindrical fibrils. Pepsin soluble collagen is produced by preparing tissue extracts with pepsin, which digests most protein structures except for the collagen triple helix. As a consequence, the nonhelical telopeptides are mostly removed by this procedure, making self-assembly more difficult. The importance of telopeptide interactions has been further demonstrated by Prockop and Fertala[59] who showed that exogenous peptides corresponding to N- or C-telopeptides could inhibit fibril formation.

As a more physiological alternative to the reconstitution of collagen fibrils from acid solutions, fibrils can also be formed in vitro from procollagen, in physiological buffers, using the metalloproteases ADAMTS-2 and BMP-1 to cleave the N- and C-propeptides, respectively. As the C-propeptide region is largely responsible for the relatively high solubility of the procollagen molecule, it is convenient to use a partially processed form with the N-propeptides removed, called pC-collagen and trigger fibril formation by the addition of BMP-1. When this is done, the initially formed (unipolar) fibrils have pointed tips, with a sharp N-terminus and a blunt C-terminus, as shown by scanning transmission electron microscopy.[60] Unlike fibrils reconstituted from acid-soluble collagen however, these de novo fibrils then go on to acquire pointed tips at both ends. This is due to the formation of bipolar fibrils in which molecules have their N-termini pointing towards the nearest tip, with an abrupt change in molecular orientation somewhere in the middle. It was subsequently shown that such bi-polar fibrils are relatively common in vivo, indicating that de novo fibril formation in vitro can reproduce at least some of the features of the in vivo system.[56]

Procollagen processing has a dramatic influence on assembly. When only the C-propeptide region is removed, the processing intermediate pN-collagen assembles to forms D-periodic sheet like structures, several microns wide but only 8 nm thick.[61] Persistence of the N-propeptide region therefore seems to sterically block lateral aggregation in one dimension. In vivo, defective N-proteinase activity is associated with a hereditary disorder originally found in cattle and sheep (dermatosparaxis) and later found in humans (Ehlers-Danlos VIIc) characterized by extreme skin fragility. When examined by electron microscopy, the collagen fibrils are no longer approximately cylindrical as in normal tissue, but instead are thin and branched. Such structures can be reproduced in vitro when the ratio of pN-collagen to collagen is high.[61] In contrast, as the proportion of collagen increases, fibril cross-sections become increasingly circular.

Another important factor that influences fibril assembly is the formation of copolymers consisting of different collagen types. Such heterotypic fibrils are the norm in most connective tissues, for example collagens I, III and V in skin, collagens I and V in cornea and collagens II and XI in cartilage. N-terminal processing of procollagens III, V and XI is slow or partial, thereby impeding fibril growth, as shown both in vitro[62-64] and in vivo,[65-67] probably by a steric blocking mechanism.[38] In contrast, the complete absence of a minor fibrillar collagen can also have dramatic effects, as has been shown most dramatically in the case of collagen V, where complete deficiency of the proα1(V) chain (which results in death at embryonic day 10) leads to the absence of procollagen V molecules and an almost total lack of collagen fibrils in the dermis, despite their being normal levels of collagen I.[68] The minor fibrillar collagens therefore seem to act as initiators of fibril formation, during the early nucleation stage.

Interactions with small leucine rich proteoglycans (SLRPs; including decorin, fibromodulin, lumican, biglycan, keratocan and osteoglycin/mimecan) also have marked effects on collagen assembly.[69-71] In vitro studies have shown that decorin, fibromodulin and lumican, with or without GAG chains, can markedly interfere with collagen fibril formation, resulting in delayed assembly and usually reduced fibril diameters. In the case of osteoglycin/mimecan, it has recently been shown that processing by BMP-1 potentiates the inhibiting effect on fibril assembly.[72] In vivo, targeted disruption (knockout) of the decorin gene is characterized by abnormal skin fragility associated with collagen fibrils having a much broader range of diameters than in normal skin, with non uniform axial mass distributions and irregular contours resulting from fibril fusion.[73] In the case of the lumican knockout,[74] fibrils in cornea which are normally thin (~30 nm) and uniform in diameter also show a loss of diameter control and organisation resulting in opaque corneas. The fibromodulin knockout results in tendon fibrils with reduced diameters, though interpretation of these data is complicated by a compensatory up-regulation of lumican expression.[75] Finally the biglycan knockout leads to an osteoporosis like defect in bone associated with large irregular fibrils.[76]

While SLRPs have usually been found to diminish the rate of fibril formation, other extracellular matrix molecules have been found to accelerate fibril formation in vitro, including fibronectin,[77] dermatopontin,[78] tenascin X,[79] perlecan,[80] hevin[81] and cartilage oligomeric matrix protein.[82] It is likely that such interactions stabilize the initial nuclei that form during the lag phase prior to subsequent fibril growth. In cell culture, targeted disruption of fibronectin expression[83] or blocking fibronectin assembly with anti-$\alpha 5\beta 1$ integrin antibodies,[84] also prevent collagen assembly.

Finally, it is now well established that in vivo, procollagen processing, secretion and fibril formation are intimately associated with membrane enclosed compartments. Prior to secretion, both procollagen processing and early fibril formation can occur in Golgi to plasma membrane carriers. Subsequently, these carriers fuse with the plasma membrane forming fibril-containing membranous extrusions projecting from the cell surface, called fibropositors.[85-87] Such structures appear to deliver newly formed fibrils to the extracellular matrix, where they join existing fibril bundles within channels formed between neighboring cells. Early fibrils in 12-16 day chick embryo tendons have diameters of about 30 nm and lengths in the range 20-30 μm.[88] Thereafter, there are sharp increases in both length and fibril diameter, probably as a result of fusion involving tapered fibril tips.

Biomedical Applications

Not surprisingly, collagens have found numerous applications in the fields of cosmetic surgery, tissue repair, tissue engineering and regenerative medicine.[89,90] These include the use of collagens as injectable dermal fillers, as vehicles for drug delivery, in wound dressings and as surgical implants and cell scaffolds.

Sources of Collagens and Collagen Biomimetics

So far, the use of collagens for biomedical applications of collagens has been limited to the fibrillar collagen types I and II and to the nonfibrillar collagen type IV. The latter, in the form of Matrigel™, a mixture of ECM proteins (laminin, collagen IV, heparan sulfate proteoglycans and entactin) extracted from the Engelbreth-Holm-Swarm mouse sarcoma, is widely used in vitro for the study of cell growth, differentiation, adhesion and migration, as well as in vivo for studies on angiogenesis and tumor growth.[91] For collagen I, traditional tissue sources include bovine bone and hide, as well as rat tail tendon, due to their widespread availability and low cost. Collagen II is extracted from bovine cartilage, though usually this requires pepsin treatment (which removes the telopeptides) for efficient solubilization. More recently, in view of concerns relating to disease transmission, notably prions, alternative sources have become available, such as porcine or fish tissues. But the risks of possible immunogenicity and disease transmission remain and this has been the driving force behind the development of recombinant human collagens and collagen biomimetic peptides.

Recombinant human collagens, produced in various mammalian cell lines (3T3, HeLa, HT1080, 293), were first reported in the early 90's.[92-95] Mammalian cells have the advantage of containing most of the enzymatic machinery required for collagen-specific posttranslational modifications, notably prolyl hydroxylation required for molecular stability at body temperature. They also permit secretion of recombinant collagens into the medium. With the exception of human embryonic kidney 293 cells however, where production levels can reach over 10 mg/l, yields with mammalian cells are relatively low.[96] High level expression (6 mg/l) of collagens in insect cells using the baculovirus system was subsequently reported,[97] but this required cotransfection with DNA coding for both subunits of prolyl hydroxylase and the protein (procollagen III) was not secreted. A further advance was the expression of recombinant procollagen III in the yeast strain *Pichia pastoris*,[98] again requiring co-expression of prolyl hydroxylase subunits. This system was subsequently optimized by Fibrogen Inc, leading to yields of up to 1500 mg/l.[99] Again however, the recombinant (pro)collagen is not secreted and purification typically relies on pepsin extraction which removes both propeptides and most of the telopeptides, resulting in impaired fibril assembly (see above). Alternative expression systems include transgenic mouse milk,[100,101] transgenic silk worms,[102] transgenic tobacco cells[99] and plants[103] but these are not yet widely used.

The availability of recombinant human collagens not only avoids problems of immunogenicity and disease transmission, but also creates the possibility of engineering new collagens as smart materials for tissue engineering applications. This approach has been elegantly pioneered by Fertala and colleagues[104,105] who have engineered variants of recombinant human collagen II in which individual D segments (see Fig. 3) were exchanged or deleted. Surprisingly, shortening the molecule by deletion of segments 2 or 3 did not affect the ability to form D-periodically banded fibrils. Furthermore, segment swapping showed that D-periodic assembly required the presence of segments 1, 4 and 5. When used as scaffolds for chondrocytes, constructs containing multiple segments 4 were found to be superior to wild type recombinant collagen II.[104]

As an alternative to recombinant collagens, a recent development has been the use of self-assembling collagen-like peptides to make novel scaffolds for tissue engineering and regenerative medicine.[106] One approach is the use of host-guest peptides,[47,107] consisting of two $(\text{Gly-Pro-Hyp})_4$ sequences (for triple helix stability) straddling a real collagen sequence, that spontaneously assemble into collagen-like trimers. Related tri-block peptide trimers have been reported to form liquid-crystalline assemblies at high concentrations.[108] Moreover, $(\text{Pro-Hyp-Gly})_{10}$ peptides assemble into branched filamentous triple-helical networks in a temperature and concentration-dependent manner, unlike $(\text{Pro-Pro-Gly})_{10}$ triple-helices which do not.[109] Host-guest peptides (containing an integrin binding sequence) have also been constructed with an N-terminal cysteine and a C-terminal thioester.[110] When incubated in an aqueous solvent at neutral pH, these peptides undergo triple-helix formation and also spontaneous chemical ligation to form polymers 10-20 nm in diameter and several microns in length. Another approach, inspired by the notion of "sticky ends" in DNA ligation, involves the synthesis of a molecule with a triple-helical core with two of the chains having N-terminal $(\text{Pro-Y-Gly})_5$ extensions and the third a C-terminal $(\text{Pro-Y-Gly})_5$ extension (where Y = Pro or Hyp)[111] or a variant thereof.[112] Molecules then associate by self-complementary interactions to form near-continues triple helices hundreds of nm in length. Finally, synthetic hydrogels of collagen-like peptides (containing cell recognition and matrix metalloprotease degradation sites) and branched polyethylene glycol have given insights into the mechanisms of bone regeneration.[113]

Collagen-Based Biomaterials

Collagens have several advantages as biomaterials, including their biocompatibility, biodegradability, ability to add functional groups, haemostatic properties and relatively low antigenicity. Antigenicity is usually minimal for atelocollagen (i.e., when the telopeptides are removed) though this effects fibril formation (see above). Collagens can also be produced in different forms: films, foams and gels. Collagen films are usually prepared by simple air drying of acidic collagen solutions. Such films have found widespread applications in drug delivery and growth factor release.[90]

Collagen foams are prepared by freeze-drying acidic collagen solutions, thereby creating an open porous network. Such foams are frequently stabilized by the addition of other naturally occurring polymers, such as glycosaminoglycans or chitosan. More recently, another way of creating porous scaffolds is the use of electrospinning,[114] in which collagen is transferred to the target under an electric potential using organic solvents. Collagen gels are prepared as described earlier, by neutralizing the pH and raising the temperature of acidic collagen solutions. An advantage of gels is that when cells are added following the neutralization step, they can be directly incorporated into the newly formed scaffold. The purpose of a tissue engineering scaffold is to provide an environment conducive to cell growth and tissue neosynthesis that is gradually degraded over time. Various methods have been developed to control the rate of degradation, including dehydrothermal treatment, chemical cross-linking, using gluteraldehyde, diphenylphosphorylazide (DPPA) or a combination of 1-ethyl-3-(3-dimethyl aminopropyl) carbodiimide (EDC) and *N*-hydroxysuccinimide (NHS),[115] or enzymatic cross-linking with transglutaminase.[116]

Skin is the area of tissue engineering where collagen-based biomaterials are the most highly developed.[117] A number of products are now commercially available, including Integra™, a collagen/chondroitin-6-sulphate cross-linked sponge with controlled pore size, used as a dermal replacement in the repair of burn injuries and scar contractures.[118] Integra™ promotes cellular infiltration and ECM neosynthesis and is eventually replaced by autologous dermal tissue. It also includes an outer silicone layer which is eventually replaced by epidermis. Another approach is to reconstitute collagen gels in the presence of dermal fibroblasts, thereby creating a simple dermal substitute. When coated with epidermal keratinocytes, the resulting product, marketed as Apligraf™, can be used in the treatment of venous leg ulcers and diabetic foot ulcers.[119] Though Apligraf™ is made using allogenic cells (i.e., not from the recipient), it persists long enough on the host to stimulate successful healing.

Cartilage is an avascular tissue with very limited ability to repair. This has led to the development of numerous tissue engineering approaches, notably the use of collagen-based sponges and gels as cell scaffolds. In order to generate sufficient numbers of cells, such approaches require an in vitro cell expansion step. Cartilage cells (chondrocytes) are notoriously prone to losing their differentiated phenotype in 2D culture. Chondrocyte culture in collagen gels, however, has been shown to preserve the cartilage phenotype over long periods[120] and this approach has been exploited in the latest generation of cartilage repair technologies (CaReS™). Unlike cartilage, bone is highly vascularised and has a strong capacity to repair. But there is a limit to the size of the lesions that repair spontaneously and this requires the development of scaffolds that permit colonization and differentiation of osteoblasts, as well as neovascularisation. Collagen-hydroxyapatite composite scaffolds have been developed in this regard (Healos™, Collapat™, Collagraft™) and when combined with 3D-printing technology,[121] these may allow the reconstruction of the complex architectures found in bone.[122]

Tissue engineering of nerves, blood vessels, heart valves, tendons and ligaments is faced with the problem of reconstructing tissues with a high degree of fiber alignment. Several methods have been devised to align collagenous scaffolds. These include harnessing the propensity of cell containing collagen gels to spontaneously contract by tethering the gel at each end,[89,123] carrying out fibril assembly under flow conditions[124] or at high concentrations,[125] electrospinning onto a rotating target[114] or reconstituting fibrils in the presence of a strong magnetic field.[126,127] In the latter approach, alignment is perpendicular to the field and is uniform over a large area. Only newly formed fibril segments are subject to alignment, since the diamagnetic anisotropy of individual collagen molecules is too small. In addition, once formed, insoluble fibrils are not affected by the magnetic field. This makes it possible to build up stacks of aligned fibrils with different orientations, by a layer-by-layer approach, as recently used[128] to reconstruct the plywood-like structure of the corneal stroma (Fig. 4).

Figure 4. Orthogonal layers of aligned collagen fibrils used for reconstruction of human corneal stroma. Scanning electron microscopy. From reference 128 with permission.

Conclusion

The collagen superfamily has turned out to be involved in a wide diversity of different forms of assembly and biological functions, with the triple-helical motif frequently serving as a strut in large multi-domain molecules and complexes. It is becoming increasingly clear that the three-dimensional structures of collagenous assemblies, which affects not only their ability to bind ligands but also their physical and mechanical properties, plays essential roles in directing cell behavior and providing the right microenvironments for cell differentiation. The ability to produce recombinant collagens and collagen-like biomimetics, along with recent developments in scaffold technology, heralds a new era in the applications of collagens in tissue engineering and regenerative medicine.

References

1. Kadler KE, Baldock C, Bella J et al. Collagens at a glance. J Cell Sci 2007; 120:1955-1958.
2. Ricard-Blum S, Ruggiero F, van der Rest M. The collagen superfamily. Top Curr Chem 2005; 247:35-84.
3. Myllyharju J, Kivirikko KI. Collagens, modifying enzymes and their mutations in humans, flies and worms. Trends Genet 2004; 20:33-43.
4. Kadler KE. Extracellular matrix 1: fibril-forming collagens. Protein Profile 1995; 2:491-619.
5. Hulmes DJS. Building collagen molecules, fibrils and suprafibrillar structures. J Struct Biol 2002; 137:2-10.
6. Plumb DA, Dhir V, Mironov A et al. Collagen XXVII is developmentally regulated and forms thin fibrillar structures distinct from those of classical vertebrate fibrillar collagens. J Biol Chem 2007; 282:12791-12795.
7. Eyre DR, Pietka T, Weis MA et al. Covalent cross-linking of the NC1 domain of collagen type IX to collagen type II in cartilage. J Biol Chem 2004; 279:2568-2574.

8. Nishiyama T, Mcdonough AM, Bruns RR et al. Type XII and XIV collagens mediate interactions between banded collagen fibers in vitro and may modulate extracellular matrix deformability. J Biol Chem 1994; 269:28193-28199.
9. Koch M, Schulze J, Hansen U et al. A novel marker of tissue junctions: collagen XXII. J Biol Chem 2004; 279:22514-22521.
10. Ruehl M, Erben U, Schuppan D et al. The elongated first fibronectin type III domain of collagen XIV is an inducer of quiescence and differentiation in fibroblasts and preadipocytes. J Biol Chem 2005; 280:38537-38543.
11. Asamura K, Abe S, Imamura Y et al. Type IX collagen is crucial for normal hearing. Neuroscience 2005; 132:493-500.
12. Baldock C, Sherratt MJ, Shuttleworth CA et al. The supramolecular organization of collagen VI microfibrils. J Mol Biol 2003; 330:297-307.
13. Knupp C, Squire JM. Molecular packing in network-forming collagens. Adv Protein Chem 2005; 70:375-403.
14. Hudson BG, Tryggvason K, Sundaramoorthy M et al. Alport's syndrome, Goodpasture's syndrome and type IV collagen. N Engl J Med 2003; 348:2543-2556.
15. Uitto J, Christiano AM. Molecular basis for the dystrophic forms of epidermolysis bullosa: Mutations in the type VII collagen gene. Arch Dermatol Res 1994; 287:16-22.
16. Brittingham R, Uitto J, Fertala A. High-affinity binding of the NC1 domain of collagen VII to laminin 5 and collagen IV. Biochem Biophys Res Commun 2006; 343:692-699.
17. Ghai R, Waters P, Roumenina LT et al. C1q and its growing family. Immunobiology 2007; 212:253-266.
18. Stephan S, Sherratt MJ, Hodson N et al. Expression and supramolecular assembly of recombinant α1(viii) and α2(viii) collagen homotrimers. J Biol Chem 2004; 279:21469-21477.
19. Kwan APL, Cummings CE, Chapman JA et al. Macromolecular organization of chicken type X collagen in vitro. J Cell Biol 1991; 114:597-604.
20. Franzke CW, Bruckner P, Bruckner-Tuderman L. Collagenous transmembrane proteins: recent insights into biology and pathology. J Biol Chem 2005; 280:4005-4008.
21. Cui CY, Schlessinger D. EDA signaling and skin appendage development. Cell Cycle 2006; 5:2477-2483.
22. van Berkel TJ, Out R, Hoekstra M et al. Scavenger receptors: friend or foe in atherosclerosis? Curr Opin Lipidol 2005; 16:525-535.
23. Maertens B, Hopkins D, Franzke CW et al. Cleavage and oligomerization of gliomedin, a transmembrane collagen required for node of ranvier formation. J Biol Chem 2007; 282:10647-10659.
24. Holmskov U, Thiel S, Jensenius JC. Collections and ficolins: humoral lectins of the innate immune defense. Annu Rev Immunol 2003; 21:547-578.
25. Takamatsu N, Kojima M, Taniyama M et al. Expression of multiple alpha1-antitrypsin-like genes in hibernating species of the squirrel family. Gene 1997; 204:127-132.
26. Nishida M, Funahashi T, Shimomura I. Pathophysiological significance of adiponectin. Med Mol Morphol 2007; 40:55-67.
27. Schaffler A, Scholmerich J, Salzberger B. Adipose tissue as an immunological organ: Toll-like receptors, C1q/TNFs and CTRPs. Trends Immunol 2007; 28:393-399.
28. Davis JG, Oberholtzer JC, Burns FR et al. Molecular cloning and characterization of an inner ear-specific structural protein. Science 1995; 267:1031-1034.
29. Braghetta P, Ferrari A, De Gemmis P et al. Overlapping, complementary and site-specific expression pattern of genes of the EMILIN/Multimerin family. Matrix Biol 2004; 22:549-556.
30. Veit G, Kobbe B, Keene DR et al. Collagen XXVIII, a novel von Willebrand factor A domain containing protein with many imperfections in the collagenous domain. J Biol Chem 2006; 281:3494-3504.
31. Bon S, Ayon A, Leroy J et al. Trimerization domain of the collagen tail of acetylcholinesterase. Neurochem Res 2003; 28:523-535.
32. Cameron GJ, Alberts IL, Laing JH et al. Structure of type I and type III heterotypic collagen fibrils: An X-ray diffraction study. J Struct Biol 2002; 137:15-22.
33. Ottani V, Martini D, Franchi M et al. Hierarchical structures in fibrillar collagens. Micron 2002; 33:587-596.
34. Holmes DF, Gilpin CJ, Baldock C et al. Corneal collagen fibril structure in three dimensions: Structural insights into fibril assembly, mechanical properties and tissue organization. Proc Natl Acad Sci USA 2001; 98:7307-7312.
35. Orgel JP, Irving TC, Miller A et al. Microfibrillar structure of type I collagen in situ. Proc Natl Acad Sci USA 2006; 103:9001-9005.
36. Hulmes DJS, Wess TJ, Prockop DJ et al. Radial packing, order and disorder in collagen fibrils. Biophys J 1995; 68:1661-1670.

37. Smith JW. Molecular pattern in native collagen. Nature 1968; 219:157-158.
38. Holmes DF, Kadler KE. The 10 + 4 microfibril structure of thin cartilage fibrils. Proc Natl Acad Sci USA 2006; 103:17249-17254.
39. Gutsmann T, Fantner GE, Venturoni M et al. Evidence that collagen fibrils in tendons are inhomogeneously structured in a tubelike manner. Biophys J 2003; 84:2593-2598.
40. Myllyharju J. Intracellular posttranslational modifications of collagens. Top Curr Chem 2005; 247:115-247.
41. Hulmes DJS. Collagen diversity, synthesis and assembly. In: Fratzl P, ed. Collagen—Structure and Mechanics. New York: Springer, 2008:
42. Morello R, Bertin TK, Chen Y et al. CRTAP is required for prolyl 3- hydroxylation and mutations cause recessive osteogenesis imperfecta. Cell 2006; 127:291-304.
43. Cabral WA, Chang W, Barnes AM et al. Prolyl 3-hydroxylase 1 deficiency causes a recessive metabolic bone disorder resembling lethal/severe osteogenesis imperfecta. Nat Genet 2007; 39:359-365.
44. Myllyla R, Wang C, Heikkinen J et al. Expanding the lysyl hydroxylase toolbox: new insights into the localization and activities of lysyl hydroxylase 3 (LH3). J Cell Physiol 2007; 212:323-329.
45. Ruotsalainen H, Sipila L, Vapola M et al. Glycosylation catalyzed by lysyl hydroxylase 3 is essential for basement membranes. J Cell Sci 2006; 119:625-635.
46. Koide T, Nagata K. Collagen biosynthesis. Top Curr Chem 2005; 247:85-114.
47. Brodsky B, Persikov AV. Molecular structure of the collagen triple helix. Adv Protein Chem 2005; 70:301-339.
48. Ishida Y, Kubota H, Yamamoto A et al. Type I collagen in Hsp47-null cells is aggregated in endoplasmic reticulum and deficient in N-propeptide processing and fibrillogenesis. Mol Biol Cell 2006; 17:2346-2355.
49. Canty EG, Kadler KE. Procollagen trafficking, processing and fibrillogenesis. J Cell Sci 2005; 118:1341-1353.
50. Porter S, Clark IM, Kevorkian L et al. The ADAMTS metalloproteinases. Biochem J 2005; 386:15-27.
51. Hopkins DR, Keles S, Greenspan DS. The bone morphogenetic protein 1/Tolloid-like metalloproteinases. Matrix Biol 2007; 26:508-523.
52. Seidah NG, Prat A. Proprotein convertases in the secretory pathway, cytosol and extracellular milieu. Essays Biochem 2005; 38:79-94.
53. Lucero HA, Kagan HM. Lysyl oxidase: an oxidative enzyme and effector of cell function. Cell Mol Life Sci 2006; 63:2304-2316.
54. Verderio EA, Johnson TS, Griffin M. Transglutaminases in wound healing and inflammation. Prog Exp Tumor Res 2005; 38:89-114.
55. Avery NC, Bailey AJ. The effects of the Maillard reaction on the physical properties and cell interactions of collagen. Pathol Biol (Paris) 2006; 54:387-395.
56. Kadler KE, Holmes DF, Trotter JA et al. Collagen fibril formation. Biochem J 1996; 316:1-11.
57. Williams BR, Gelman RA, Poppke DC et al. Collagen fibril formation: optimal in vitro conditions and preliminary kinetic results. J Biol Chem 1978; 235:6578-6585.
58. Holmes DF, Capaldi MJ, Chapman JA. Reconstitution of collagen fibrils in vitro; the assembly process depends on the initiating procedure. Int J Biol Macromol 1986; 8:161-166.
59. Prockop DJ, Fertala A. Inhibition of the self-assembly of collagen I into fibrils with synthetic peptides—Demonstration that assembly is driven by specific binding sites on the monomers. J Biol Chem 1998; 273:15598-15604.
60. Holmes DF, Chapman JA, Prockop DJ et al. Growing tips of type-I collagen fibrils formed in vitro are near-paraboloidal in shape, implying a reciprocal relationship between accretion and diameter. Proc Nat Acad Sci USA 1992; 89:9855-9859.
61. Hulmes DJS, Kadler KE, Mould AP et al. Pleomorphism in type I collagen fibrils produced by persistence of the procollagen N-propeptide. J Mol Biol 1989; 210:337-345.
62. Romanic AM, Adachi E, Kadler KE et al. Copolymerization of pNcolagen III and collagen I. J Biol Chem 1991; 266:12703-12709.
63. Birk DE, Fitch JM, Babiarz JP et al. Collagen fibrillogenesis in vitro: interaction of types I and V collagen regulates fibril diameter. J Cell Sci 1990; 95(Pt 4):649-657.
64. Blaschke UK, Eikenberry EF, Hulmes DJS et al. Collagen XI nucleates self-assembly and limits lateral growth of cartilage fibrils. J Biol Chem 2000; 275:10370-10378.
65. Liu X, Wu H, Byrne M et al. Type III collagen is crucial for collagen I fibrillogenesis and for normal cardiovascular development. Proc Natl Acad Sci USA 1997; 94:1852-1856.
66. Wenstrup RJ, Florer JB, Davidson JM et al. Murine model of the Ehlers-Danlos syndrome. col5a1 haploinsufficiency disrupts collagen fibril assembly at multiple stages. J Biol Chem 2006; 281:12888-12895.

67. Li Y, Lacerda DA, Warman ML et al. A fibrillar collagen gene, Col11a1, is essential for skeletal morphogenesis. Cell 1995; 80:423-430.
68. Wenstrup RJ, Florer JB, Brunskill EW et al. Type V collagen controls the initiation of collagen fibril assembly. J Biol Chem 2004; 279:53331-53337.
69. Graham HK, Holmes DF, Watson RB et al. Identification of collagen fibril fusion during vertebrate tendon morphogenesis. The process relies on unipolar fibrils and is regulated by collagen-proteoglycan interaction. J Mol Biol 2000; 295:891-902.
70. Iozzo RV. The biology of the small leucine-rich proteoglycans—Functional network of interactive proteins. J Biol Chem 1999; 274:18843-18846.
71. McEwan PA, Scott PG, Bishop PN et al. Structural correlations in the family of small leucine-rich repeat proteins and proteoglycans. J Struct Biol 2006; 155:294-305.
72. Ge G, Seo NS, Liang X et al. Bone morphogenetic protein-1/tolloid-related metalloproteinases process osteoglycin and enhance its ability to regulate collagen fibrillogenesis. J Biol Chem 2004; 279:41626-41633.
73. Danielson KG, Baribault H, Holmes DF et al. Targeted disruption of decorin leads to abnormal collagen fibril morphology and skin fragility. J Cell Biol 1997; 136:729-743.
74. Chakravarti S, Magnuson T, Lass JH et al. Lumican regulates collagen fibril assembly: Skin fragility and corneal opacity in the absence of lumican. J Cell Biol 1998; 141:1277-1286.
75. Svensson L, Aszùdi A, Reinholt FP et al. Fibromodulin-null mice have abnormal collagen fibrils, tissue organization and altered lumican deposition in tendon. J Biol Chem 1999; 274:9636-9647.
76. Xu T, Bianco P, Fisher LW et al. Targeted disruption of the biglycan gene leads to an osteoporosis-like phenotype in mice. Nat Genet 1998; 20:78-82.
77. Speranza ML, Valentini G, Calligaro A. Influence of fibronectin on the fibrillogenesis of type I and type III collagen. Coll Relat Res 1987; 7:115-123.
78. MacBeath JR, Shackleton DR, Hulmes DJS. Tyrosine-rich acidic matrix protein (TRAMP) accelerates collagen fibril formation in vitro. J Biol Chem 1993; 268:19826-19832.
79. Minamitani T, Ikuta T, Saito Y et al. Modulation of collagen fibrillogenesis by tenascin-X and type VI collagen. Exp Cell Res 2004; 298:305-315.
80. Kvist AJ, Johnson AE, Morgelin M et al. Chondroitin sulfate perlecan enhances collagen fibril formation. Implications for perlecan chondrodysplasias. J Biol Chem 2006; 281:33127-33139.
81. Sullivan MM, Barker TH, Funk SE et al. Matricellular hevin regulates decorin production and collagen assembly. J Biol Chem 2006; 281:27621-27632.
82. Halasz K, Kassner A, Morgelin M et al. COMP acts as a catalyst in collagen fibrillogenesis. J Biol Chem 2007; 282:31166-31173.
83. Velling T, Risteli J, Wennerberg K et al. Polymerization of type I and III collagens is dependent on fibronectin and enhanced by integrins alpha 11beta 1 and alpha 2beta 1. J Biol Chem 2002; 277:37377-37381.
84. Li S, Van Den DC, D'Souza SJ et al. Vascular smooth muscle cells orchestrate the assembly of type I collagen via alpha2beta1 integrin, RhoA and fibronectin polymerization. Am J Pathol 2003; 163:1045-1056.
85. Birk DE, Trelstad RL. Extracellular compartments in tendon morphogenesis: collagen fibril, bundle and macroaggregate formation. J Cell Biol 1986; 103:231-240.
86. Birk DE, Zycband EI, Winkelmann DA et al. Collagen fibrillogenesis in situ: fibril segments are intermediates in matrix assembly. Proc Natl Acad Sci USA 1989; 86:4549-4553.
87. Canty EG, Lu Y, Meadows RS et al. Coalignment of plasma membrane channels and protrusions (fibripositors) specifies the parallelism of tendon. J Cell Biol 2004; 165:553-563.
88. Birk DE, Nurminskaya MV, Zycband EI. Collagen fibrillogenesis in situ: Fibril segments undergo post depositional modifications resulting in linear and lateral growth during matrix development. Develop Dynam 1995; 202:229-243.
89. Brown RA, Phillips JB. Cell responses to biomimetic protein scaffolds used in tissue repair and engineering. Int Rev Cytol 2007; 262:75-150.
90. Lee CH, Singla A, Lee Y. Biomedical applications of collagen. Int J Pharm 2001; 221:1-22.
91. Kleinman HK, Martin GR. Matrigel: basement membrane matrix with biological activity. Semin Cancer Biol 2005; 15:378-386.
92. Ala-Kokko L, Hyland J, Smith C et al. Expression of a human cartilage procollagen gene (COL2A1) in mouse 3T3 cells. J Biol Chem 1991; 266:14175-14178.
93. Mazzorana M, Gruffat H, Sergeant A et al. Mechanisms of Collagen Trimer Formation—Construction and Expression of a Recombinant Minigene in HeLa Cells Reveals a Direct Effect of Prolyl Hydroxylation on Chain Assembly of Type- XII Collagen. J Biol Chem 1993; 268:3029-3032.
94. Fertala A, Sieron AL, Ganguly A et al. Synthesis of recombinant human procollagen II in a stably transfected tumour cell line (Ht1080). Biochem J 1994; 298:31-37.

95. Tillet E, Wiedemann H, Golbik R et al. Recombinant expression and structural and binding properties of alpha 1 (VI) and alpha 2 (VI) chains of human collagen type VI(1994; 221:177). Eur J Biochem 1994; 222:1064.
96. Bulleid NJ, John DCA, Kadler KE. Recombinant expression systems for the production of collagen. Biochem Soc Trans 2000; 28:350-353.
97. Lamberg A, Helaakoski T, Myllyharju J et al. Characterization of human type III collagen expressed in a baculovirus system—Production of a protein with a stable triple helix requires coexpression with the two types of recombinant prolyl 4-hydroxylase subunit. J Biol Chem 1996; 271:11988-11995.
98. Vuorela A, Myllyharju J, Nissi R et al. Assembly of human prolyl 4-hydroxylase and type III collagen in the yeast Pichia pastoris: formation of a stable enzyme tetramer requires coexpression with collagen and assembly of a stable collagen requires coexpression with prolyl 4-hydroxylase. EMBO J 1997; 16:6702-6712.
99. Baez J, Olsen D, Polarek JW. Recombinant microbial systems for the production of human collagen and gelatin. Appl Microbiol Biotechnol 2005; 69:245-252.
100. John DCA, Watson R, Kind AJ et al. Expression of an engineered form of recombinant procollagen in mouse milk. Nat Biotechnol 1999; 17:385-389.
101. Toman PD, Pieper F, Sakai N et al. Production of recombinant human type I procollagen homotrimer in the mammary gland of transgenic mice. Transgenic Res 1999; 8:415-427.
102. Tomita M, Munetsuna H, Sato T et al. Transgenic silkworms produce recombinant human type III procollagen in cocoons. Nature Biotechnol 2003; 21:52-56.
103. Merle C, Perret S, Lacour T et al. Hydroxylated human homotrimeric collagen I in Agrobacterium tumefaciens-me diated transient expression and in transgenic tobacco plant. FEBS Lett 2002; 515:114-118.
104. Ito H, Steplewski A, Alabyeva T et al. Testing the utility of rationally engineered recombinant collagen-like proteins for applications in tissue engineering. J Biomed Mater Res A 2006; 76:551-560.
105. Steplewski A, Hintze V, Fertala A. Molecular basis of organization of collagen fibrils. J Struct Biol 2007; 157:297-307.
106. Koide T. Triple helical collagen-like peptides: engineering and applications in matrix biology. Connect Tissue Res 2005; 46:131-141.
107. Fields GB, Lauer JL, Dori Y et al. Protein-like molecular architecture: biomaterial applications for inducing cellular receptor binding and signal transduction. Biopolymers 1998; 47:143-151.
108. Martin R, Waldmann L, Kaplan DL. Supramolecular assembly of collagen triblock peptides. Biopolymers 2003; 70:435-444.
109. Kar K, Amin P, Bryan MA et al. Self-association of collagen triple helix peptides into higher order structures. J Biol Chem 2006; 281:33283-33290.
110. Paramonov SE, Gauba V, Hartgerink JD. Synthesis of collagen-like peptide polymers by native chemical ligation. Macromolecules 2005; 38:7555-7561.
111. Kotch FW, Raines RT. Self-assembly of synthetic collagen triple helices. Proc Natl Acad Sci USA 2006; 103:3028-3033.
112. Koide T, Homma DL, Asada S et al. Self-complementary peptides for the formation of collagen-like triple helical supramolecules. Bioorg Med Chem Lett 2005; 15:5230-5233.
113. Lutolf MP, Hubbell JA. Synthetic biomaterials as instructive extracellular microenvironments for morphogenesis in tissue engineering. Nat Biotechnol 2005; 23:47-55.
114. Matthews JA, Wnek GE, Simpson DG et al. Electrospinning of collagen nanofibers. Biomacromolecules 2002; 3:232-238.
115. Pieper JS, Hafmans T, Veerkamp JH et al. Development of tailor-made collagen-glycosaminoglycan matrices: EDC/NHS crosslinking and ultrastructural aspects. Biomaterials 2000; 21:581-593.
116. Orban JM, Wilson LB, Kofroth JA et al. Crosslinking of collagen gels by transglutaminase. J Biomed Mater Res A 2004; 68:756-762.
117. Metcalfe AD, Ferguson MW. Tissue engineering of replacement skin: the crossroads of biomaterials, wound healing, embryonic development, stem cells and regeneration. J R Soc Interface 2007; 4:413-437.
118. Jones I, Currie L, Martin R. A guide to biological skin substitutes. Br J Plast Surg 2002; 55:185-193.
119. Ehrenreich M, Ruszczak Z. Update on tissue-engineered biological dressings. Tissue Eng 2006; 12:2407-2424.
120. Galois L, Hutasse S, Cortial D et al. Bovine chondrocyte behaviour in three-dimensional type I collagen gel in terms of gel contraction, proliferation and gene expression. Biomaterials 2006; 27:79-90.
121. Mironov V, Boland T, Trusk T et al. Organ printing: computer-aided jet-based 3D tissue engineering. Trends Biotechnol 2003; 21:157-161.
122. Wahl DA, Czernuszka JT. Collagen-hydroxyapatite composites for hard tissue repair. Eur Cell Mater 2006; 11:43-56.

123. Shi Y, Rittman L, Vesely I. Novel geometries for tissue-engineered tendonous collagen constructs. Tissue Eng 2006; 12:2601-2609.
124. Elsdale T, Bard J. Collagen substrata for studies on cell behavior. J Cell Biol 1972; 54:626-637.
125. Mosser G, Anglo A, Helary C et al. Dense tissue-like collagen matrices formed in cell-free conditions. Matrix Biol 2006; 25:3-13.
126. Torbet J, Ronziere M-C. Magnetic alignment of collagen during self-assembly. Biochem 1984; 219:1057-1059.
127. Ceballos D, Navarro X, Dubey N et al. Magnetically aligned collagen gel filling a collagen nerve guide improves peripheral nerve regeneration. Exp Neurol 1999; 158:290-300.
128. Torbet J, Malbouyres M, Builles N et al. Orthogonal scaffold of magnetically aligned collagen lamellae for corneal stroma reconstruction. Biomaterials 2007; 28:4268-4276.
129. Beck K, Brodsky B. Supercoiled protein motifs: The collagen triple-helix and the alpha-helical coiled coil. J Struct Biol 1998; 122:17-29.

Chapter 3

Short-Order Tendons:
Liquid Crystal Mesophases, Metal-Complexation and Protein Gradients in the Externalized Collagens of Mussel Byssal Threads

Matthew J. Harrington* and J. Herbert Waite

Abstract

Mussel byssal threads are externally secreted collagneous protein fibers that tether marine mussels to hard surfaces in the rocky intertidal zone and prevent dislodgement from the drag and lift forces generated by crashing waves. At low strain values, byssal threads mechanically resemble tendon with a stiffness ~1 GPa and a high resilience above 90%. However, unlike tendon, threads do not rupture at 12% strain, but instead undergo yield and continue extending until twice their initial length. In doing so, they dissipate up to 70% of the applied wave energy. Byssal threads are also able to self-heal after yield-induced stress softening, recovering initial material properties in a time dependent fashion. The added mechanical features of byssal threads appear to be an adaptation for functioning outside the organism in the tumultuous intertidal zone. Since both tendon and byssal threads are composed of over 50% collagen, it is intriguing that their mechanical properties would be so varied. This review focuses on the structure-property relationships that exist between the unique mechanical performance of the byssal threads and the biochemistry of the modified collagens (preCols) of which they are composed. In particular, we discuss the role of protein liquid crystal phases, histidine-metal complexation and graded distribution of preCol variants in the assembly process and mechanical function of mussel byssal threads.

Introduction

Collagen is a ubiquitous protein component of bone and load-bearing connective tissues such as tendon, cartilage and skin in many organisms (see Chapter 2). Over 20 types of collagen have been identified in vertebrates[1] and many additional types occur in invertebrates. Although collagens exhibit extraordinary diversity, all proteins or protein domains regarded as collagenous are characterized by a primary sequence triplet repeat Gly-X_1-X_2, in which X_1 and X_2 are most commonly proline and the proline in the X_2 position is almost always hydroxylated.[2] For instance, type I collagen, which is the most common variant, has over 300 consecutive repeats of the Gly-X_1-X_2 triplet.[1] Each collagen polypeptide chain assumes a left-handed polyproline II helix (3_1). Three of these left-handed helices further assemble into a right-handed triple helix in which the glycine residues, owing to their low steric interference, are oriented towards the inside of the triple helix.[2]

Most collagens, such as types I-III, form fibrillar structures and function in tissues that are wholly enclosed within the body of the organism. However, there are several unusual collagens that

*Corresponding Author: Matthew J. Harrington—Department of Molecular, Cellular and Developmental Biology (MCDB), University of California, Santa Barbara, Santa Barbara, California 93106, USA. Email: mjh@umail.ucsb.edu

Fibrous Proteins, edited by Thomas Scheibel. ©2008 Landes Bioscience.

function outside the body of the organism. Examples include gooseberry sawfly silk,[3] gorgonian coral,[4] dogfish egg cases,[5] Peripatus slime[6] and the focus of this review, mussel byssal threads.[7]

Collagenous materials functioning outside an organism are more apt to encounter harsh and unpredictable environmental conditions than those situated within the controlled environment of the body. Because of this, externalized collagens have many interesting adaptations that distinguish them from typical vertebrate collagens. Many of these adaptations mirror how the material copes with environmental damage and degradation. For instance, below ~10% strain, the mechanical performance of mussel byssal threads and animal tendon is nearly identical, reflecting their similar composition. However, unlike tendon, byssal threads yield at a sub-critical strain prior to catastrophic failure. This gives them an ultimate strain of over 100% and a maximum hysteresis of around 70% vs 13% and 10% respectively for tendon collagen.[8] Byssal threads are also distinct from tendon in being able to autonomically self-heal without the aid of cellular repair processes.[9] As will be discussed further on in this review, it is the thread's ability to yield and dissipate energy, as well as to recover initial material properties postyield that allows the mussel to remain securely attached in the chaotic environment of the marine intertidal zone. The molecular structure of the collagenous proteins that make up the byssal thread reflect these adapted mechanical functions and by comparing the structure of byssal collagen with tendon collagen, we can speculate intelligently on structure-property relationships in byssal threads.

The focus of this chapter is on how the mussel byssal thread, an externally functioning collagenous biomaterial, has adapted the molecular structure, chemistry and assembly of its protein components in order to deal with the pressures encountered in the turbulent intertidal zone. Specifically, we will discuss how marine mussels utilize liquid crystal mesophases, metal-coordination chemistry and graded distributions of protein variants in the production, mechanical function and maintenance of byssal threads. The design paradigms distilled from fundamental research on the molecular nature of byssal threads could have important implications for the engineering of fibrous polymers.

Background

For most of their life history, marine mussels of the species *Mytilus* remain securely attached to hard surfaces within the intertidal zone.[10] Here, crashing waves, which can reach local velocities of up to 25 m/s, impose considerable lift and drag forces.[11] Mussels hold tenaciously to the surface with an array of completely acellular, collagenous threads known collectively as a byssus (Fig. 1), which is able to dissipate up to 70% of the applied wave energy.[9,12] Byssal threads are secreted one

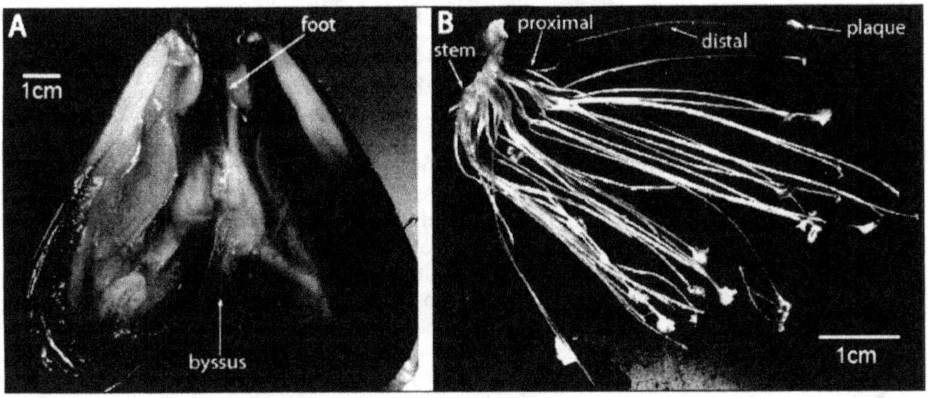

Figure 1. View of a shucked mussel with its byssus still attached. A) A specimen of *Mytilus californianus* with severed adductor muscles revealing the foot (thread forming organ) and attached byssus. B) A more detailed photograph of a byssus dissected from the mussel with the different regions denoted.

at a time by the mussel foot, an extendable organ in the mussel, tongue-like in appearance, in which the soluble protein precursors that make up the threads are synthesized and stored prior to thread formation (Fig. 2).[13] When a mussel wishes to fashion a new thread, it sticks its foot out from its shell, finds an appropriate spot for attachment and secretes the soluble protein into a groove running

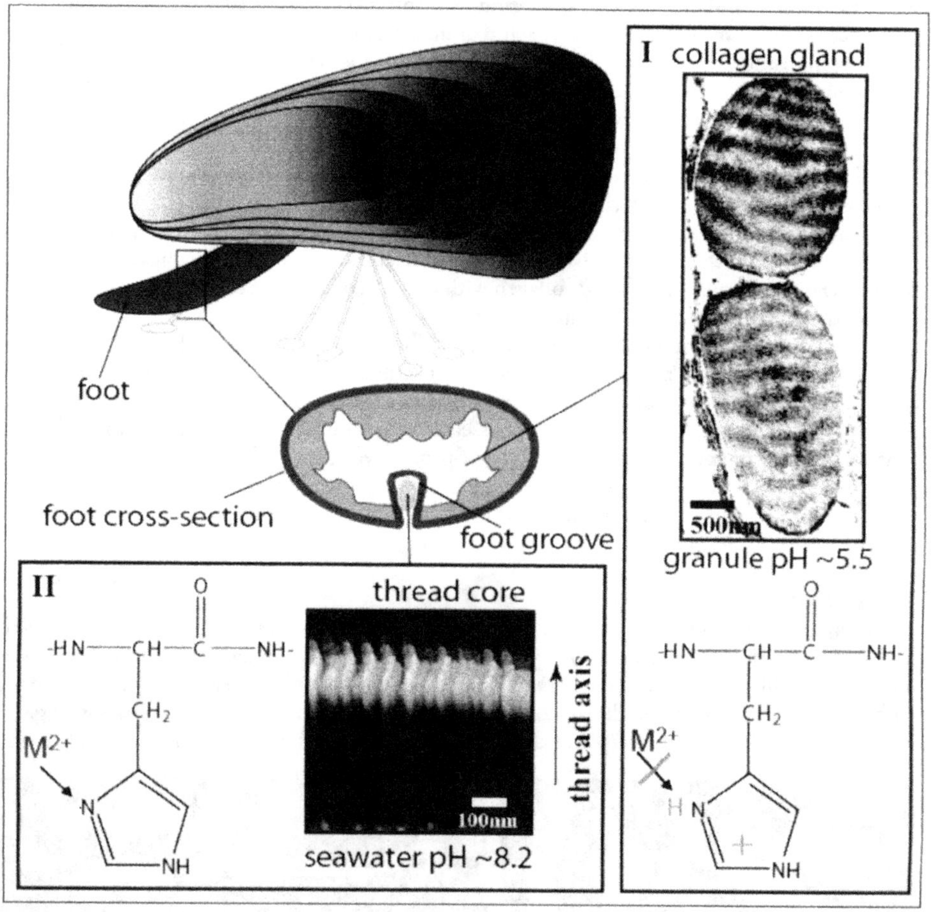

Figure 2. A scheme outlining critical steps in the thread forming process. Byssal threads are formed by the mussel foot when the soluble protein precursors are secreted into the foot groove. I) The proteins, known as preCols, are stored as a smectic liquid crystal phase in holocrine secretory granules in specialized glands of the foot as seen by TEM. The smectic banding pattern within the granules indicates layering of preCols with the dark bands representing the rod-like collagen domain and the light bands representing the amorphous end regions. The pH of the granules is ~5.5 at which the histidine residues at the termini of preCol are protonated and unable to participate as ligands in metal chelate cross-links. II) After secretion into the foot groove where the pH is closer to 8, the His residues are deprotonated allowing them to cross-link and lock the smectic organization in place as seen in AFM images of the thread core. AFM image reprinted with permission from reference 29.

along the length of the ventral side of the foot.[14] Muscular contractions around the groove shape the secreted protein into a thread, the whole process taking 3-5 minutes.[14]

Each byssal thread has four distinct regions: the stem, the plaque, the proximal thread and the distal thread, all of which are encased by a separately secreted outer coating (Fig. 3). The stem attaches the thread to the soft tissue of the mussel, but relatively little is known about the composition or mechanical performance of this region compared to the others. The plaque is an adhesive pad with which the thread adheres to hard surfaces within the intertidal, making it an interesting model system for studying the nature of underwater adhesives.[15] Many of the adhesive proteins composing the plaque have been characterized[15] and dihydroxyphenylalanine (DOPA), an uncommon posttranslational modification of tyrosine is a major component of all of them, suggesting that it may play an important role in adhesion.[16]

Between the stem and plaque is the thread itself, which in mytilids is about 2-5 cm in length, between 100-200 microns in diameter and is divided morphologically and mechanically into the proximal and distal regions (Fig. 3). The distal region, furthest from the organism, is characterized by a stiffness comparable to that of tendon, an extensibility of around 100%, a large hysteresis and the ability to self-heal.[9] The proximal region, which is closer to the organism, is much less stiff, but twice as extensible when compared to the distal region.[17] The transition between the two portions of the thread appears to be gradual, but the distal region makes up about 80 of the total length.[17] Table 1 indicates how the two regions of the byssal thread compare mechanically with other common fibrous protein materials as well as several engineered fibrous materials.

The collagenous proteins of byssus are not directly exposed to the marine milieu, but are completely enclosed by a cuticle 2-5 μm thick. The cuticle has a granular structure, exhibits a hydrated hardness that is six times greater than the core and is dominated by a single protein known as mfp-1.[18] A recent comparison of byssal threads with granular and nongranular cuticles showed that granular structures enabled breaking strains more than twice those of nongranular cuticles.[18] This has important implications for the wear resistance and half-life of byssal threads on mussels in the wave-swept intertidal zone where thread strains are likely to be large and to occur with regularity.

Byssal Thread Composition

Wide angle X-ray scattering (WAXS) of the distal portion of byssal threads from *Mytilus sp* provided the first convincing evidence for the presence of collagen within threads.[19,20] A β-sheet like structure within the distal region of the thread and reminiscent of silks was also suggested. Further support for the presence of collagen was ascertained by amino acid analysis of hydrolyzed threads, which revealed a composition rich in glycine, proline and hydroxyproline.[21,22] Although traditional collagen extraction procedures, such as acid solubilization, proved ineffective with byssal threads, extensive pepsinization of threads released several pepsin-resistant polypeptide fragments.[23] Partial Edman sequence of the fragments revealed collagen-like sequences, consistent with the resistance of the collagen triple helix to pepsin degradation.[23] The partial peptide sequences were used to design oligonucleotides to screen a cDNA library generated from the mussel foot and the full cDNA-deduced sequence of the first byssal collagen revealed a novel modified collagen.[24]

Three byssal collagens have been characterized in *Mytilus edulis* and denoted preCol D,[24] preCol NG,[25] and preCol P.[26] Homologous preCol variants were discovered and sequenced in two related species, *Mytilus galloprovincialis*[27] and *Mytilus californianus*.[28] Byssal threads are 95% protein by dry weight and the combined preCols represent 96% of the protein content in the distal region of the thread and 66% in the proximal region of the thread.[12] Not only is the percentage composition of total preCol different in the two regions of the thread, but it was also shown that the three preCol variants are not uniformly distributed along the length of the thread; that is, preCol D and P exist in complementary distal-to-proximal gradients (Fig. 3).[23]

The structure of each preCol variant resembles a block copolymer in which the clearly defined modular blocks or domains resemble known load-bearing proteins.[12] Each of the three variants has

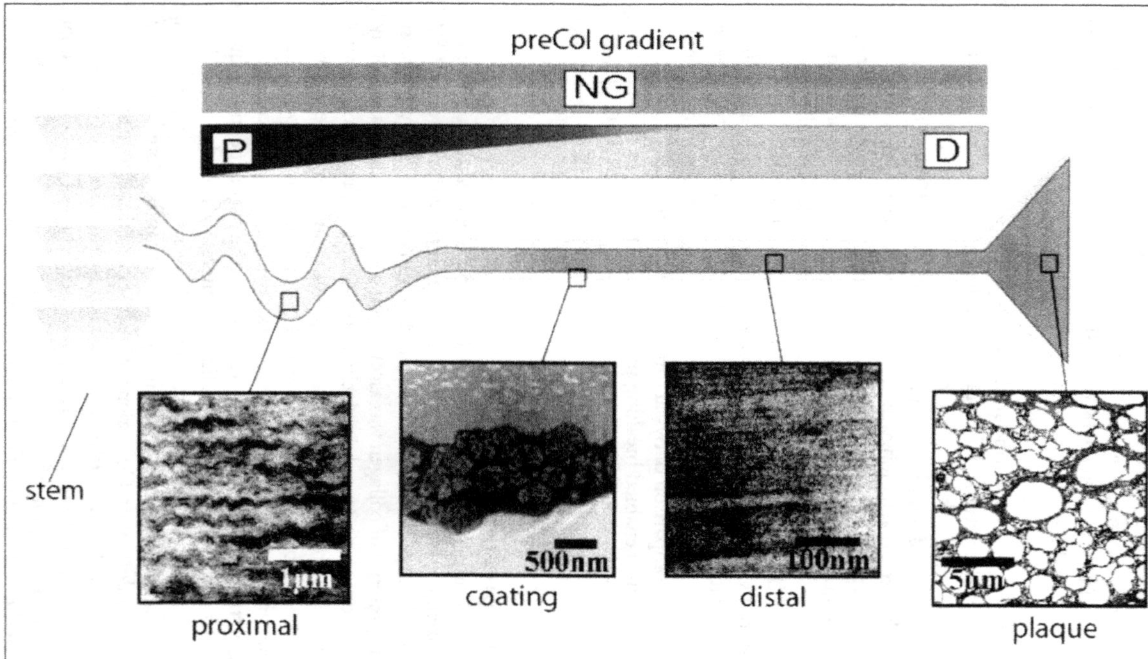

Figure 3. A schematic view of a longitudinal section through an isolated byssal thread showing important structural features. Representative TEM images of the different regions of a thread highlight the changes in morphology from the soft tissue of the organism to the rocky surface of the intertidal. Graded morphological and mechanical differences from the distal to proximal region of the thread correlate closely with a gradient in the relative composition of preCol variants; preCol-D, -NG and -P.

Table 1. Mechanical properties of biological and synthetic materials in tension

Material	Modulus (Gpa)	Strength (Gpa)	Extensibility (mm/mm)	Hysteresis (%)	Toughness (MJ m-3)
Distal Thread (*M.californianus*)[17]	0.87	0.075	1.09	72	45
Collagen (mammalian tendon)[65]	1.2	0.12	0.13	10	6
Dragline silk (*A. diadematus*)[66,67]	10	1.1	0.3	65	160
Kevlar[68]	130	3.6	0.027	nd	50
Proximal Thread (*M.californianus*)[17]	0.016	0.035	2	47	35
Elastin (bovine ligament)[69]	0.001	0.002	1.5	10	1.6
Viscid silk (*A. diadematus*)[66,67]	0.003	0.5	2.7	65	150

a central kinked collagen domain, variable flanking domains at either end of the collagen domain and histidine-rich regions at both termini (Fig. 4).

The central collagen domain is the largest region with between 140 and 180 repeats of the canonical Gly-X_1-X_2 collagen triplet sequence in each variant. However, these repeats are interrupted 1-4 times by single amino acid substitutions, deletions, or short inserts that break up the repeat structure.[12,29] Aberrations in the repeat structure of collagen are believed to cause bends or kinks in the normally rod-like structure of the triple helix as seen in dogfish egg case collagen[5] and the C1q heterotrimer.[30] Based on X-ray crystallography of short, synthetic collagen peptides containing similar aberrations, the break is believed to maintain collagens in full lateral register rather than in the quarter stagger typical of fibrillar collagen.[31]

The flanking domains are adjacent to the collagen domain on both the N and C terminal sides and differ significantly between the three variants. PreCol D, which is dominant at the distal end of the thread and decreases in abundance approaching the proximal end, has flanking domains that resemble a motif sequence of spider dragline silk with many polyalanine runs separated by glycine-rich spacer regions (see Chapter 9). PreCol P, which exists in a gradient complementary to preCol D, has flanking regions that closely resemble elastin and preCol NG, which is uniformly distributed throughout the thread, has flanking domains that resemble Gly-rich plant cell wall proteins.[7]

Based on the similarity of these flanking domains to known load-bearing proteins and the respective stiffnesses of these proteins, it is predicted that the preCol variants will have very different mechanical properties.[12] For instance, all other domains being similar between variants, preCol D, which has flanking domains resembling a stiff silk, can be expected to have a higher overall stiffness than preCol P with its elastin-like flanking domains. PreCol NG is predicted to have a stiffness intermediate to preCol D and P. As mentioned, it has been demonstrated that preCol P and D exist in a gradient along the thread axis with D prominent in the distal region and P prominent in the proximal region.[23] This gradient correlates very closely to the mechanical gradient from proximal to distal and it has been proposed that they are intimately associated.[12] However, it must be stressed that the proximal region is only 66% preCol and the effect of the other 34% protein making up the proximal region cannot be discounted. The implications of this gradient will be discussed in detail later in the chapter.

Overall, preCol compositions having only 2-4 mol% histidine can hardly be considered histidine-rich, however, the overwhelming majority of histidine residues are concentrated within the N and C-terminal 50 residues thus elevating their local abundance to ~20 mol%.[28] Histidine residues are well known to participate as reversible ligands in coordination complexes with divalent transition metal ions such as zinc(II) and copper(II).[32] Since atomic absorption spectrophotometry[33] and PIXE analysis[34] have both shown significant concentrations of the transition metals Zn, Cu and Fe in the thread, the histidine residues in these terminal domains are believed to play a pivotal role as cross-linking ligands in the thread.[12] The His-rich terminal domains of all three variants also

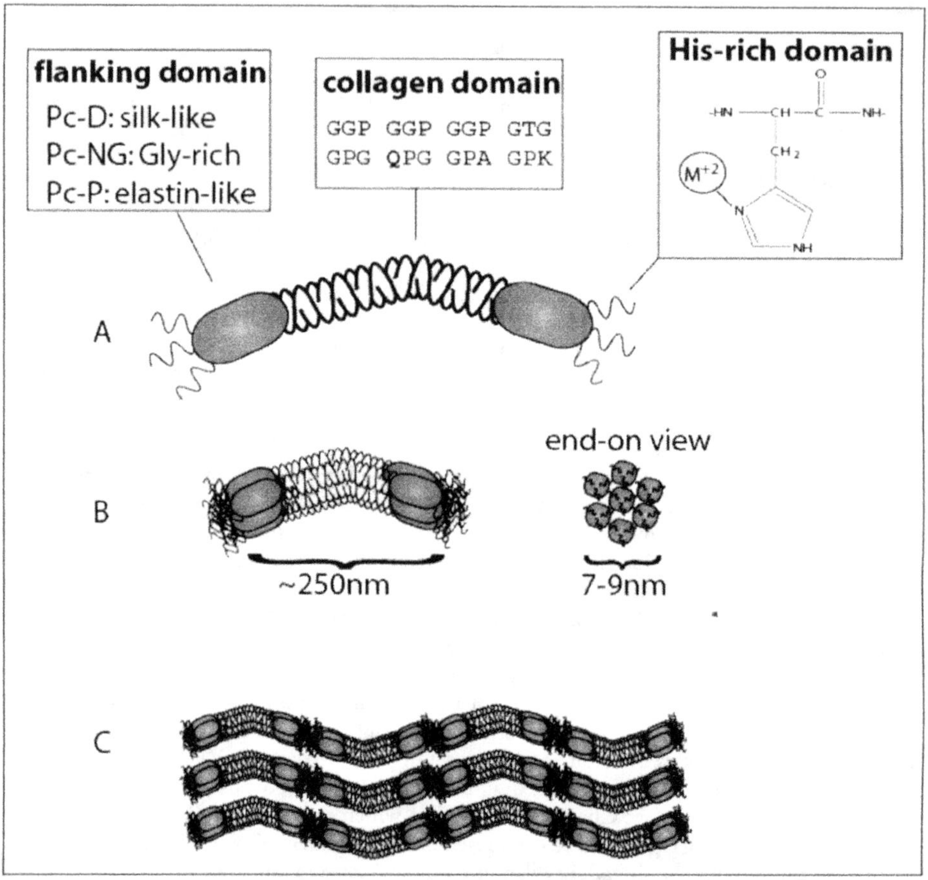

Figure 4. A schematic view of preCol structure and assembly. A) Three variants of preCol have been identified in *Mytilus* species; preCol-D, -NG and -P. All three have a common block copolymer-like structure in which the blocks resemble the sequence of known load-bearing proteins including a central collagen domain, variable flanking domains and His-rich domains. PreCols form triple helices because of the central collagen domain. The flanking domains are the most different between the three variants ranging from silk-like to elastin-like. The His-rich domains are believed to participate as ligands in metal chelate complexes at neutral to basic pH. B) PreCol triple helices are believed to further assemble into 6 + 1 hexagonal bundles based on evidence from TEM of collagen granules[13] and AFM of thread cores.[29] C) Within the granules, preCol bundles are arranged into smectic liquid crystal phase with C2 symmetry. This organization is retained in mature threads.

contain between two and four highly conserved residues of tyrosine at both ends that are known or proposed to be posttranslationally modified to dihydroxyphenylalanine (DOPA).[24] The highly reactive side chain of the oxidized form of DOPA could form covalent cross-links between the ends of neighboring preCols.[35] The potential role of histidine-metal and DOPA cross-links in the formation and mechanical function of threads will be amplified later.

The remaining proteins (dry weight) that have been identified are the thread matrix protein (TMP-1) in both the distal and proximal portion of the thread and the proximal thread matrix protein (PTMP-1) in the proximal portion.[36,37] These may play crucial roles in the assembly and mechanics of byssal threads, but it is too early to speculate.

Byssal Thread Synthesis

Byssus formation is a carefully choreographed holocrine secretion and specific glands within the foot of the mussel are responsible for producing the protein constituents of the different components of the byssal thread, including the phenol gland (plaque), the accessory gland (coating) and the collagen gland (thread core).[38] The collagen gland surrounds the foot groove proximal to the distal depression (Fig. 2). Cells within the gland synthesize and package preCol into ellipsoid granules that measure about 1.5 μm on the long axis and are stored near the groove ready for secretion (Fig. 2).[13,38] Within each granule, preCols are aligned into an ordered array resembling a smectic liquid crystal phase in which the collagen domains are aligned in layers like bottles on a shelf as shown in Figure 2 and Figure 4C.[13,29] Collagen liquid crystal phases are not unprecedented. The helical orientation of collagen in bone strongly resembles a solidified cholesteric liquid crystal phase[39] and purified type I collagen forms liquid crystal phases in vitro.[40]

The contents of the collagen granules are secreted into the foot groove during thread formation in a process resembling polymer injection molding and threads are shaped from the soluble precursor by muscular contractions in the foot.[14] The entire process of thread formation typically lasts under five minutes, allowing the mussel to respond quickly to the need for more threads to maintain a secure attachment. The smectic morphology of the granular byssal collagen is retained in the mature thread (Fig. 2),[29] however, due to extensive cross-linking, the fluidity of the protein is lost in favor of a solid tensile element. As mentioned, breaks in the triplet repeat sequence of collagen direct in-register lateral alignment of the collagen rather than the typical quarter stagger seen in fibrillar collagens.[31] The presence of aberrations in the collagen sequence of the preCols is highly conserved between species and it is likely that they are playing a similar role and directing smectic alignment.

The use of a liquid crystal phase during fiber processing has been characterized in the silk glands of spiders[41] and silk worms[42] and is utilized in the manufacturing process of Kevlar[43]. These are all examples of lyotropic liquid crystals, in which the liquid crystal forming molecule, or mesogen, is dissolved in a solvent and formation of a liquid crystal phase, or mesophase, is dependent on the mesogen concentration during solvent evaporation. When the concentration of the mesogen increases to a critical point, it becomes more energetically favorable for the elongated molecules to be aligned anisotropically than to be disordered. The lyotropic liquid crystal phase is beneficial in processing fibers for several reasons. First, contrary to intuition, the bulk viscosity of the mesophase decreases when the concentration of mesogen reaches the critical concentration, reducing the shear stress and requiring less energy to transport and process the protein.[44] Second, there is inherent order in the mesophase, yet it flows like a liquid making it ideal for injection molding and other methods of forming anisotropic fibers. In particular, the preordering of the monomer allows for simpler processing and better alignment in the final product and this in turn allows a stiffer and stronger fiber since tensile stress goes directly into stretching the polymer (protein) backbone and distorting bond angles.[45]

Mechanical Function

It is inherently difficult to precisely define the function of a biological device given the various evolutionary pressures that have optimized a particular biomaterial for a number of multifunctional roles. With regard to byssus, we focus on tenacity which can be thought of as the mechanical function of byssal threads to prevent detachment from hard surfaces and ultimately the demise of the organism. Since the distal and proximal regions of the thread are mechanically distinct, we will first discuss them individually and then will elaborate on how they work together in nature.

At low strains, the distal region of the thread has a stiffness comparable to tendon (~850 MPa for *M. californianus*) and is essentially elastic, with a resilience above 90% (Table 1).[9] At approximately 15% strain, the distal region undergoes a yield event in which the stiffness drops to almost 0 MPa, as the thread extends to about 50% strain with little resistance (Fig. 5A). After the yield region, the stiffness again increases considerably from ~50% strain until the thread breaks at about 100% strain. Yield serves the purpose of preventing catastrophic failure through dissipa-

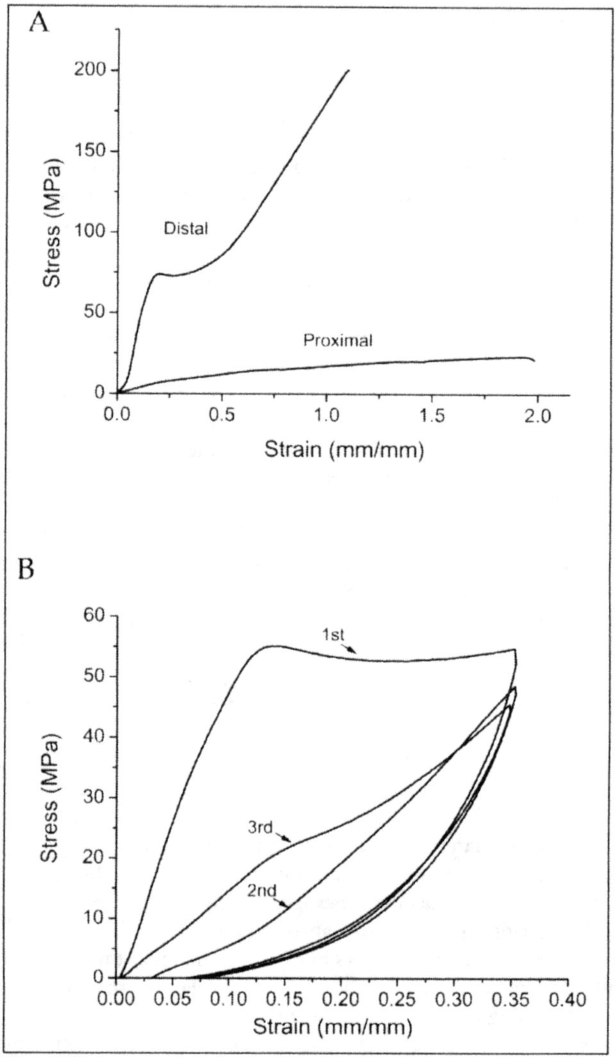

Figure 5. Mechanical performance of the different regions of mussel byssal threads. A) Stress-strain curves of the proximal and distal portions of byssal threads from *M. californianus* pulled to failure. Despite their extensive mechanical differences, it is important to note that in nature they are assembled and function in series. B) Stress-strain curve of single distal portion of a thread cycled three times illustrating its propensity to self-heal. Cycle 1 is followed immediately by cycle 2, whereas there is a 30 min rest between cycle 2 and cycle 3.

tion of applied wave energy, endowing the thread with a hysteresis of up to ~70%. This is in stark contrast to animal tendon, which has a maximum hysteresis of only 10% and fails catastrophically by 13% strain.[8] However, this is desirable in tendon since its function is to transfer energy and not to dissipate it as in the byssus.

When most common materials are taken beyond yield, it signifies the end of their functional utility since the initial modulus in subsequent cycles will be compromised by stress softening.[46] The distal region of the byssal thread also exhibits the reduction in stiffness typical of stress softening, however, the initial mechanical properties are recoverable through time dependent self-healing

(Fig. 5B).[9] Self-healing is defined as the capacity of a material to regain functionality after damage by utilizing the inherent resources of the material system in an autonomic fashion and can be measured as the healing efficiency (η),[47] which is represented as:

η = (material property healed/material property virgin) × 100%

In the case of the distal region of the thread, the healing efficiency has been measured using the strain energy to monitor recovery.[9] The healing efficiency has been shown to be both time- and strain-dependent, with efficiency increasing in proportion to the duration of rest time and with lower strain values. After a 35% strain cycle, the healing efficiency reaches 70% after 1 hour of rest based on recovery of strain energy. It is important to distinguish the acellular self-healing of the byssal threads with the healing of damaged tendon. The repair of damaged tendon involves a complicated cellular response and takes months before the tendon is even partially healed.[48] The self-healing of byssal threads, on the other hand, is completely intrinsic to the material itself and must therefore be a byproduct of the molecular composition of the threads.

At the proximal end of the thread, the mechanical properties are markedly different from the distal region. The proximal region has a stiffness about fifty times lower than the distal in *M. californianus* (15 MPa) and no obvious yield point (Fig. 5A).[17] The breaking stress is also much lower for the proximal region. However, nonreversible strain-stiffening that can double the stiffness is observed when the proximal region of the thread is agitated in aerated water prior to tensile testing.[49] Although much less stiff and strong, the proximal region is twice as extensible as the distal region and is able to extend to twice its initial length before it breaks, approaching the breaking strain of the most extensible protein fibers, such as elastin, resilin and flagelliform (viscid) silk (Table 1) (see also Chapter 9).

Of course in nature, these two distinct regions of the thread do not function independently, but are coordinated and perform mechanically in series. In the context of the whole thread, the strength of the distal region is over-designed since mechanical failure occurs primarily at the plaque or proximal region and the high breaking stress of the distal region is never reached.[17] This raises the question of the biological relevance of this over-design. Bell and Gosline[17] propose that yield is the most important aspect of the distal region and the high strength is a byproduct of designing a collagenous material that yields. Yield not only offers hysteresis for individual threads, but also allows the threads that do yield to transfer their load to the other threads in the byssus, reducing the load any one thread must bear on its own.[17] From a biological standpoint this strategy is very interesting, but the unique material properties of the distinct thread portions also merit study from a strictly biomimetic/biomaterials perspective as well.

As already mentioned above, the mechanical transition from the distal to the proximal region of the thread is not sharp, but gradual.[12] The proposed utility of this gradual transition in stiffness is to solve the problem of modulus mismatch inherent in attaching the soft mussel tissue to a hard rock in the intertidal seashore.[12] When two different materials with disparate stiffnesses and similar Poisson ratios make contact, damage to the softer partner will be directly proportional to the magnitude of the difference in stiffnesses.[50] Biology often moderates modulus mismatch arising from necessary contacts between hard and soft materials by the use of chemical or architectural gradients with mechanical consequences. The dentino-enamel junction,[51] polychaete jaws[52] and arthropod cuticle[53] are well studied examples. The centimeter-scale of the gradients in mussel byssal thread make it an attractive model for scrutinizing biological strategies for assembling functionally graded polymers.[52,54]

Structure-Function Relationships in Byssal Threads

Fundamental research in the field of biomimetics and biomaterials aims to understand how nature solves common engineering problems given that evolution has spent eons honing and refining the materials utilized by biological organisms. At the heart of this aim, from a biochemical perspective, is gaining an understanding of the structure-property relationships that exist between the design of the molecular components at the nanoscale up through the hierarchical levels of organization and the macroscopic mechanical performance of the material. Essentially, we are interested in how the basic components and their organization lead to the observable properties of a material and its function. Byssal threads are an ideal model for this sort of approach since the structural and functional similarity and divergence of the preCols from typical fibrillar collagens permits speculation on the roles that the different domains of the preCol are playing in the mechanical function of the threads. Here we consider how differences in the molecular structure of preCols when contrasted with type I collagen reveal carefully tailored structural and mechanical properties that are well suited to their function in byssal threads. In particular, we will focus on potential roles of the flanking domains and the His-rich domains within the distal region since it is composed of almost 96% preCol.

Although the specific context of their function is significantly different, tendon and byssus share some parallels: both face the common challenge of connecting a soft material (muscle in both cases) to a hard material (bone vs rocks). Interfaces between tendon and other tissues are known as entheses and involve distributions of many molecules that are still under investigation.[55] The mussel takes a different approach by fine-tuning the distribution of preCol variants to engineer a fifty-fold stiffness gradient ranging from about 800 MPa near the plaque (in *M. californianus*) to 15 MPa in the proximal portion.[17] Evidence for the role of the flanking domains in defining the mechanical stiffness of the thread comes from a study in which preCol sequences and mechanical properties of byssal threads from three related mussel species were compared.[28] The flanking domains of preCol D in the species with a 2-3-fold stiffer distal region were found to contain an increased number of polyalanine runs. Given the tendency of polyalanine domains to form beta sheet structures during shearing,[42] an increased density of polyalanines and beta structure is often suggestive of a stiffer resulting fiber. Indeed, the recombinatory replacement of glycine-rich sequences with cassettes of polyalanine has been proposed to be a driver for stiffening in spider silk evolution.[56] Since the collagen domain of preCol D in all three species is almost completely conserved, this suggests that it is the flanking domains that play the most important role in determining the overall stiffness of the distal thread.

Strain energy, or the integrated area under the stress-strain curve of a byssal thread pulled to break, is often used as a measure of fiber toughness. Byssal threads are considerably tougher than tendon mainly due to their ability to yield prior to catastrophic failure. It has been proposed that yield and toughness in the byssus are due to the breaking of noncovalent, sacrificial bonds.[28,57] The most probable candidates for sacrificial bonds in the distal region of byssal threads are histidine-metal coordination complexes and the globular flanking domains of preCol NG.[28] The flanking domains of smectic arrays of preCol NG have been shown by AFM to extend during strain whereas the collagen domains extend relatively little.[29] There is a strong resemblance of the flanking domains of preCol NG to Gly-rich plant proteins and Gly-loops, which are believed to exhibit unraveling during tensile loading and entropic recovery when the load is removed.[58] However, at this point the specific role of the Gly-rich domains of preCol NG in the mechanical performance of byssal threads is mostly speculative.

Histidine residues in the His-rich domains of the all three preCols are highly conserved between all three species that have been sequenced, suggesting an important functional role.[28] A common role of histidine in a number of proteins is as a ligand for metal coordination complexes.[59] Levels of transition metals are elevated in the thread over seawater and appear to be actively concentrated by the mussel,[33] which supports this possibility. In further support of this prediction, threads treated at various pHs below pH 8 and then mechanically tested, exhibited a sigmoidal relationship between pH and thread stiffness: as pH increased, so did stiffness, with a midpoint at pH 6.6 (Fig. 6). This correlation is very suggestive of histidine-metal complexation since complexation is only possible when a nitrogen on the imidazole ring is deprotonated. This deprotonation event has a typical pK_a ~6.5,[60] and so it is significant that the midpoint in stiffness during these tests also occurs in this range.

If threads treated at low pH are re-equilibrated at pH 8, the loss in stiffness is completely recovered, which is consistent with the reversible nature of the coordination complex. However, if threads are treated with diethylpyrocarbonate (DEPC) at low pH and re-equilibrated to pH 8,

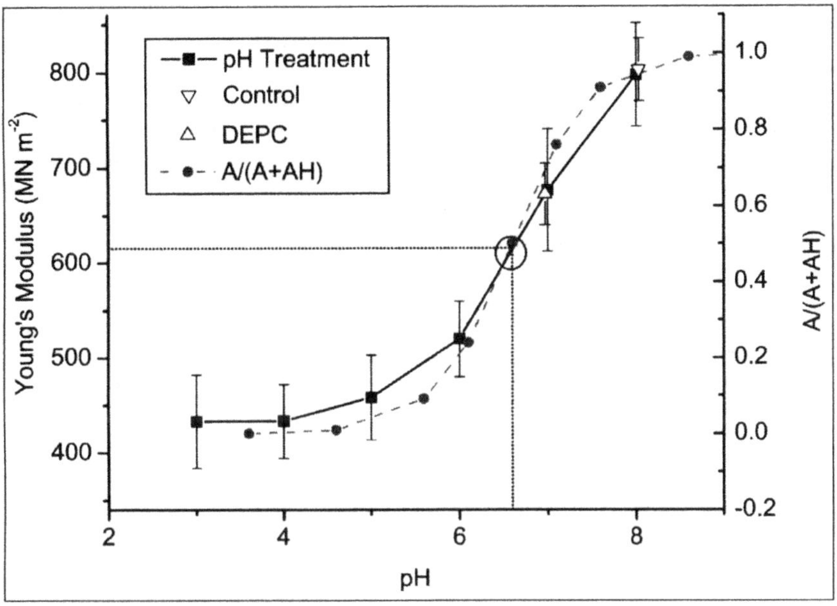

Figure 6. Initial modulus (mean ± s.e.m.) of distal byssal threads plotted against the pH at which they were incubated. Dotted line and circle denotes the midway point of the Young's modulus and the corresponding pH of 6.6. The titration of a nitrogen proton on the histidine imidazole ring (pK_a 6.6) is plotted in a light gray dashed line to illustrate that the modulus curve closely follows the characteristic sigmoidal shape, suggesting that histidine protonation may influence thread mechanics. The inverted triangle represents the initial stiffness of threads treated at pH 5.5 and brought back to pH 8 and tested, indicating that pH-induced softening is reversible. The triangle represents the initial stiffness of threads treated in DEPC at pH 5.5, brought back to pH 8 and then tested. DEPC susceptibility further supports the role of His in the mechanical properties of threads. Reprinted with permission from reference 28.

the stiffness is only partially recovered (Fig. 6). Since DEPC is known to prevent His from binding metal by specifically derivatizing one of the nitrogens on the imidazole ring of histidine at low concentrations,[61] the role of histidine as an important ligand in byssal thread cross-linking is further bolstered. Previous suggestion of metal ion participation in the mechanical performance of the distal portion comes from reference 34, showing that threads exhaustively treated with a broad spectrum metal chelator such as EDTA exhibit a significant loss in initial modulus and lose the ability to yield and self heal. From an environmental perspective, the pH-dependent mechanical performance of byssal threads is somewhat troubling since ocean pH has been dropping, presumably due to increased carbon dioxide in the atmosphere.[62] This could have potentially drastic effects on the ability of mussels to survive in high intertidal zones and lead to enhanced shoreline erosion.

These studies seem to indicate that His-metal coordination complexes are important cross-links within the byssal thread and contribute to the initial stiffness of the thread. Since these bonds are weaker than covalent bonds, each byssal thread could dissipate wave energy in the form of molecular friction (coordinate bonds breaking) rather than catastrophically rupturing the covalent bonds of the peptide backbone of the preCol. Compared to traditional tendon collagen, this allows for a higher ultimate strain, increased toughness and significant increase in hysteresis during cyclic tensile testing. Rupture of the coordinate cross-links results in a loss of stiffness in subsequent cycles (stress softening) since they had previously resisted deformation during tensile strain. However, the stiffness, strain energy and slack of byssal threads pulled past yield are recoverable in a time-dependent self-healing process. We propose that self-healing of the thread is linked to the reformation of the His-metal sacrificial bonds which has been demonstrated between Ni, Zn, Cu, Co and model His-rich peptides over thousands of cycles.[32] True healing, however, requires more than the reformation of His metal bonds; the bonding ligands and metal ions must be returned to their prestrained positions. The reversal of strain occurring during material deformation is usually associated with elastic recoil. Since different species exhibit different rates of healing[9] and all share highly conserved His-rich domains, a variable elastomeric component is predicted to occur in parallel with the His-metal bonds. The flanking domains of preCol-NG could provide that elasticity but this remains to be demonstrated.[28] A molecular model depicting a possible scenario of the reversible yield behavior in the distal region of byssal threads has been proposed; [28] the details of which can be seen in Figure 7.

The pH sensitivity of His-metal coordinate cross-links in byssal threads may also provide a convenient switch for rapidly locking the smectic liquid phase during preCol secretion in thread formation. Most regulated secretory granules have a slightly acidic pH around 5.5.[63] At pH ≤ 5.5, most histidine residues will be almost completely protonated with a +1 charge leading to mutual

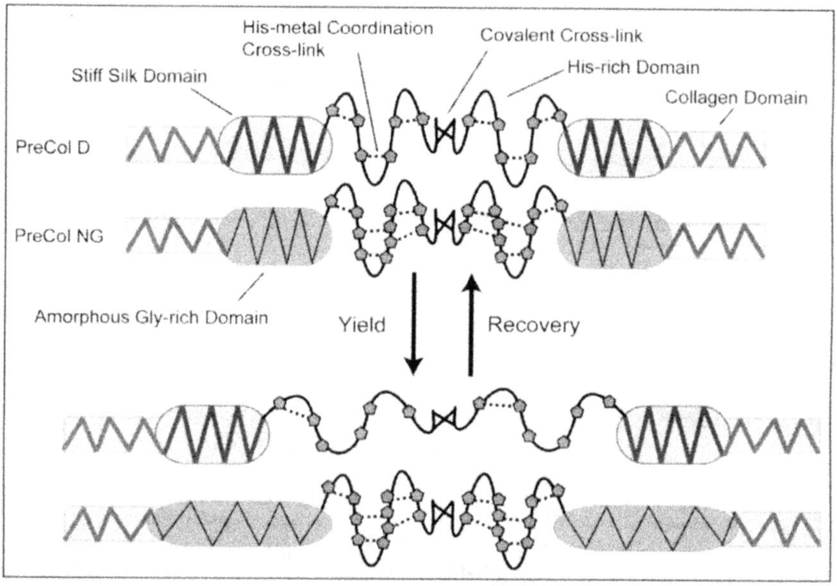

Figure 7. Molecular model of yield and self-healing in the distal byssal thread. Data suggest that His-metal coordinate cross-links in the termini of the preCols are reversibly sacrificial bonds in yield and self-healing. A difference in healing rate between species, despite high homology of the His-rich domains, suggests that there is a separate entropic elastomeric component driving recovery. Since preCol D and NG are colocalized in the distal thread and since preCols are believed to align in register, we propose that the flanking domain of NG is playing this role. In this model, the stiffness of the amorphous Gly-rich flanking domain of preCol NG is roughly the same as the His-rich domain of preCol D and the silk domain and NG His-rich domain are somewhat stiffer. During yield, the His-metal cross-links of preCol D begin to rupture as the Gly-rich domain begins to unravel, whereas the stiffer silk domain, collagen domains and NG His-rich domain stay folded. When the tensile force is released, there is a time-dependent entropic drive for the Gly-rich domain of preCol NG to recover its initial length. In doing so, the histidine residues are brought back within proximity of one another allowing the metal coordination bonds to reform and initial mechanical properties to be recovered. DiDOPA cross-links are proposed to be the covalent bonds between the ends of preCols in series. Reprinted with permission from reference 28.

repulsion and no metal binding (Fig. 2). Under these conditions, associations favoring a liquid crystal phase will dominate. However, once the granules are ruptured and their contents released into the mussel foot groove (pH ~8), the histidines undergo deprotonation and become ligands in His-metal cross-links, locking the smectic phase in place (Fig. 2). It is likely that this is only the first step in cross-linking the thread and more permanent covalent cross-links are subsequently formed. This supposition is supported by the fact that threads become much more brittle and darker over their lifetimes[9,49] suggesting that tanning (DOPA oxidation) is occurring and ultimately leading to formation of covalent adducts with other amino acids such as DOPA (diDOPA)[35] and cysteine (5-cysteinyl-DOPA).[64]

Conclusions

Byssal thread biochemistry, mechanical properties and architecture are adaptively linked to the mussel's survival in high surf habitats. Research on byssal threads in the last decade has focused on making connections between the molecular design of the modified collagens that compose them with their assembly and mechanical performance. The design paradigms distilled from this research may be harnessed by materials engineers to incorporate certain desirable properties into the design of fibrous polymers. Such properties include engineered yield and self-healing, pH-triggered locking of mesophases and graded distribution of block compolymer variants with different mechanical properties. In particular, we believe that the ligands of metal chelate cross-links present themselves as multifunctional chemical moities that could be introduced into a material as a pH triggered polymerization agent and as a means of toughening a material with reversible sacrificial bonds. Of course, as we have learned from studying byssal threads, the arrangement and organization of these bonds is of the utmost importance to function. Current and future research on byssal threads is focused on further understanding the nature of the molecular structure-mechanical property relationships discussed in this review because only with a thorough fundamental understanding can we begin to apply what has been learned.

Acknowledgements

We acknowledge N. Holten-Andersen (coating TEM) and X. Qin (distal, proximal, granule TEM) for providing images for figures in the chapter.

References

1. Ricard-Blum S, Ruggiero F, van der Rest M. The collagen superfamily. In: Brinckmann J, Notbohm H, Müller PK, eds. Topics in Current Chemistry. Berlin Heidelberg: Springer-Verlag, 2005:35-84.
2. van der Rest M, Garrone R. Collagen family of proteins. FASEB J 1991; 5(13):2814-2823.
3. Spiro RG, Lucas F, Rudall KM. Glycosylation of hydroxylysine in collagens. Nat New Biol 1971; 231(19):54-55.
4. Kingsley RJ, Tsuzak M, Watabe N et al. Collagen in the spicule organic matrix of the gorgonian Leptogorgia virgulata. Biol Bull 1990; 179:207-213.
5. Knight DP, Hunt S. Fibril structure of collagen in egg capsule of dogfish. Nature 1974; 249(5455):379-380.
6. Benkendorff K, Beardmore K, Gooley AA et al. Characterisation of the slime gland secretion from the peripatus, Euperipatoides kanangrensis (Onychophora: Peripatopsidae). Comp Biochem Physiol B Biochem Mol Biol 1999; 124(4):457-465.
7. Waite JH, Qin X-X, Coyne KJ. The peculiar collagens of mussel byssus. Matrix Biol 1998; 17(2):93-106.
8. Gosline J, Lillie M, Carrington E et al. Elastic proteins: biological roles and mechanical properties. Philos Trans R Soc Lond B Biol Sci 2002; 357:121-132.
9. Carrington E, Gosline J. Mechanical design of mussel byssus: Load cycle and strain rate dependence. American Malacological Bulletin 2004; 18:135-142.
10. Yonge CM. On the primitive significance of the byssus in the bivalvia and its effects in evolution. J Mar Biol 1962; 42:113-125.
11. Denny M, Gaylord B. The mechanics of wave-swept algae. J Exp Biol 2002; 205(10):1355-1362.
12. Waite JH, Vaccaro E, Sun CJ et al. Elastomeric gradients: a hedge against stress concentration in marine holdfasts? Philos Trans R Soc Lond B Biol Sci 2002; 357(1418):143-153.

13. Zuccarello LV. The collagen gland of Mytilus galloprovincialis: An ultrastructural and cytochemical study on secretory granules. J Ultrastruct Res 1980; 73:135-147.
14. Waite JH. The formation of mussel byssus: anatomy of a natural manufacturing process. Results Probl Cell Differ 1992; 19:27-54.
15. Waite JH. Adhesion a la moule. Integr Comp Biol 2002; 42(6):1172-1180.
16. Lee H, Scherer NF, Messersmith PB. Single-molecule mechanics of mussel adhesion. Proc Natl Acad Sci 2006; 103(35):12999-13003.
17. Bell E, Gosline J. Mechanical design of mussel byssus: material yield enhances attachment strength. J Exp Biol 1996; 199(4):1005-1017.
18. Holten-Andersen N, Fantner GE, Hohlbauch S et al. Protective coatings on extensible biofibers. Nat Mater 2007; 6(9):669-672.
19. Mercer EH. Observations on the molecular structure of byssus fibers. Australian Journal of Marine Freshwater Research 1952; 3(2):199-205.
20. Rudall KM. The distribution of collagen and chitin. Symp Soc Exp Biol 1955; 9:49-71.
21. Pujol JP, Rolland M, Lasry S et al. Comparative study of amino acid composition of byssus in some common bivalve molluscs. Comp Biochem Physiol 1970; 34(1):193-201.
22. Bdolah A, Keller PJ. Isolation of collagen granules from foot of sea mussel, Mytilus californianus. Comp Biochem Physiol B Biochem Mol Biol 1976; 55(2):171-174.
23. Qin X-X, Waite JH. Exotic collagen gradients in the byssus of the mussel Mytilus edulis. J Exp Biol 1995; 198(3):633-644.
24. Qin X-X, Coyne KJ, Waite JH. Tough Tendons: Mussel byssus has collagen with silk-like domains. J Biol Chem 1997; 272(51):32623.
25. Qin X-X, Waite JH. A potential mediator of collagenous block copolymer gradients in mussel byssal threads. Proc Natl Acad Sci 1998; 95(18):10517-10522.
26. Coyne KJ, Qin X-X, Waite JH. Extensible collagen in mussel byssus: A natural block copolymer. Science 1997; 277(5333):1830-1832.
27. Lucas JM, Vaccaro E, Waite JH. A molecular, morphometric and mechanical comparison of the structural elements of byssus from Mytilus edulis and Mytilus galloprovincialis. J Exp Biol 2002; 205(12):1807-1817.
28. Harrington MJ, Waite JH. Holdfast heroics: comparing the molecular and mechanical properties of Mytilus californianus byssal threads. J Exp Biol 2007: 210:4307.
29. Hassenkam T, Gutsmann T, Hansma P et al. Giant bent-core mesogens in the thread forming process of marine mussels. Biomacromolecules 2004; 5(4):1351-1354.
30. Kilchherr E, Hofmann H, Steigemann W et al. Structural model of the collagen-like region of C1q comprising the kink region and the fiber-like packing of the six triple helices 1985; 186(2):403-415.
31. Bella J, Liu J, Kramer RZ et al. Conformational effects of Gly-X-Gly Interruptions in the collagen triple helix. J Mol Biol 2006; 362:298-311.
32. Schmitt L, Ludwig M, Gaub HE et al. A metal-chelating microscopy tip as a new toolbox for single-molecule experiments by atomic force microscopy. Biophys J 2000; 78(6):3275-3285.
33. Coombs TL, Keller PJ. Mytilus byssal threads as an environmental marker for metals. Aquatic Toxicology 1981; 1(5-6):291-300.
34. Vaccaro E, Waite JH. Yield and postyield behavior of mussel byssal thread: A self-healing biomolecular material. Biomacromolecules 2001; 2(3):906-911.
35. McDowell LM, Burzio LA, Waite JH et al. Rotational echo double resonance detection of cross-links formed in mussel byssus under high-flow Stress. J Biol Chem 1999; 274(29):20293–20295.
36. Sun C, Lucas JM, Waite JH. Collagen-binding matrix proteins from elastomeric extraorganismic byssal fibers. Biomacromolecules 2002; 3(6):1240-1248.
37. Sagert J, Sun CJ, Waite JH. Chemical subtlties of mussel and polychaete holdfasts. In: Smith AM, Callow JA, eds. Biological Adhesives. Berlin Heidelberg: Springer-Verlag, 2006:125-140.
38. Tamarin A, Keller PJ. Ultrastructural study of byssal thread forming system in Mytilus. J Ultrastruc Res 1972; 40(3-4):401-416.
39. Giraud-Guille MM. Plywood structures in nature. Current Opinion in Solid State and Materials Science 1998; 3(3):221-227.
40. Gobeaux F, Belamie E, Mosser G et al. Cooperative ordering of collagen triple helices in the dense state. Langmuir 2007; 23(11):6411-6417.
41. Vollrath F, Knight DP. Liquid crystalline spinning of spider silk. Nature 2001; 410(6828):541-548.
42. Asakura T, Umemura K, Nakzawa Y et al. Some observations on the structure and function of the spinning apparatus in the silkworm Bombyx mori. Biomacromolecules 2007; 8:175-181.
43. Tanner D. The Kevlar innovation. R&D Innovator 1995; 4:131-147.
44. Chae HG, Kumar S. Rigid-rod polymeric fibers. Journal of Applied Polymer Science 2006; 100:791-802.

45. Wang X-J, Zhou Q-F. Liquid Crystalline Polymers. Singapore: World Scientific Publishing Co, 2004.
46. Vincent J. Structural Biomaterials. Princeton: Princeton University Press, 1982.
47. Kessler MR, Sottos NR, White SR. Self-healing structural composite materials. Composites: Part A 2003; 34:743-753.
48. Butler DL, Juncosa N, Dressler MR. Functional efficacy of tendon repair processes. Annu Rev Biomed Eng 2004; 6:303-29.
49. Sun CJ, Vaccaro E, Waite JH. Oxidative stress and the mechanical properties of naturally occurring chimeric collagen-containing fibers. Biophys J 2001; 81(6):3590-3595.
50. Suresh S, Mortensen A. Fundamentals of functionally graded materials. London: IOM communications Ltd; 1998.
51. Imbeni V, Kruzic JJ, Marshall GW et al. The dentin—enamel junction and the fracture of human teeth. Nat Mater 2005; 4:229-232.
52. Waite JH, Lichtenegger HC, Stucky GD et al. Exploring molecular and mechanical gradients in structural bioscaffolds. Biochemistry 2004; 43(24):7653-7662.
53. Raabe D, Sachs C, Romano P. The crustacean exoskeleton as an example of a structurally and mechanically graded biological nanocomposite material. Acta Materialia 2005; 53:4281-4292.
54. van Hest JCM, Tirrell DA. Protein-based materials, toward a new level of structural control. Chem Commun (Camb) 2001; 19:1897-1904.
55. Benjamin M, Toumi H, Ralphs JR et al. Where tendons and ligaments meet bone: attachment sites ('entheses') in relation to exercise and/or mechanical load. J Anat 2006; 208(4):471-490.
56. Hayashi CY, Lewis RV. Evidence from flagelliform silk cDNA for the structural basis of elasticity and modularnature of spider silks. J Mol Biol 1998; 275:773-784.
57. Bertoldi K, Boyce MC. Mechanics of the hysteretic large strain behavior of mussel byssus threads. J Mat Sci 2007; 42(21):8943-8956.
58. Sachetto-Martins G, Franco LO, de Oliveira DE. Plant glycine-rich proteins: a family or just proteins with a common motif. Biochim Biophys Acta 2000; 1492:1-14.
59. Maret W. Exploring the zinc proteome. Journal of Analytical Atomic Spectrometry 2004; 19:15-19.
60. Sundberg RJ, Martin RB. Interactions of histidine and other imidazole derivatives with transition metal ions in chemical and biological systems. Chem Rev 1974; 74(4):471-517.
61. Lundblad RL. Chemical Reagents for Protein Modification. 3rd ed. Boca Raton, Fl: CRC Press, 2005.
62. Orr JC, Fabry VJ, Aumont O et al. Anthropogenic ocean acidification over the twenty-first century and its impact on calcifying organisms. Nature 2005; 437:681-686.
63. Mellman I, Fuchs R, Helenius A. Acidification of the endocytic and exocytic pathways. Annu Rev Biochem 1986; 55:663-700.
64. Zhao H, Waite JH. Coating proteins: Structure and cross-linking in fp-1 from the green shell mussel Perna canaliculus. Biochemistry 2005; 44(48):15915-15923.
65. Pollock M, Shadwick RE. Relationship between body-mass and biomechanical properties of limb tendons in adult mammals. Amer J Phys 1994; 266(3):R1016-R1021.
66. Denny M. Physical properties of spiders silk and their role in design of orb-webs. J Exp Biol 1976; 65(2):483-506.
67. Gosline JM, Guerette PA, Ortlepp CS et al. The mechanical design of spider silks: from fibroin sequence to mechanical function. J Exp Biol 1999; 202(23):3295.
68. Gordon JE. The Science of Structures and Materials. New York: Scientific American Library, 1988.
69. Aaron BB, Gosline JM. Elastin as a random-network elastomer—A mechanical and optical analysis of single elastin fibers. Biopolymers 1981; 20(6):1247-1260.

CHAPTER 4

The Formation of Fibrin Clots

Russell F. Doolittle*

Abstract

The fibrin blood clot is composed of polymerized units of fibrinogen that are only slightly modified by thrombin-mediated proteolysis. Nonetheless, the transformation has many subtle aspects and many factors can influence the character of the mature fibrin clot. Although many details of the process are understood, some aspects remain controversial and even mysterious.

Introduction

The central events that occur during the conversion of fibrinogen to fibrin have been appreciated for more than half a century.[1]

$$F \xrightarrow{T} f \rightleftarrows f_n \longrightarrow \text{Fibrin}$$

Thrombin (T) attacks fibrinogen (F) and removes small peptides—the fibrinopeptides and as a result the parent molecules (f) spontaneously polymerize by way of a staggered overlapping process to yield two-molecule thick protofibrils (f_n). The protofibrils themselves self-assemble into longer and thicker polymers that are the strands of the fibrin gel.[1] In the more than half a century since this scenario was outlined, tens of thousands of experiments have been conducted in an effort to understand its structural and kinetic details. That the simple scheme shown above needed embellishing was quickly appreciated when it was shown that, under physiological conditions, fibrin becomes covalently cross-linked by factor XIII, a thrombin-activated enzyme subsequently shown to be a transglutaminase.[2]

Much of the early foundational work on the structure of fibrinogen was dependent on classical physico-chemical studies (Table 1); electron microscopy was also an indispensable tool throughout the history of these studies.[3-5] Among the major structural milestones that were achieved in that half century interval should be included, first the determining of the primary structure[6-8] and then, much later, an atomic-level model of the starting fibrinogen molecule determined by X-ray crystallography.[9-17] Fortuitously, a crystal structure was also obtained for a large fragment isolated from factor XIII-cross-linked fibrin[11] and as a result a glimpse was gotten of the nature of the interactions that occur between adjacent molecules in a clot. A great deal remains mysterious, however and it is these unknown or disputed aspects that will be emphasized in this chapter.

Structure of Fibrinogen

The general structure of fibrinogen follows a similar design in all vertebrates, three pairs of non-identical polypeptide chains[18] being interlinked by an extensive system of disulfide bonds.[19] Were it any other protein, the subunit structure would be designated simply $\alpha_2\beta_2\gamma_2$, but most workers in the field insist on referring to fibrinogen as $A\alpha_2B\beta_2\gamma_2$, the A and the B referring to the presence of the fibrinopeptides. The overall molecular weights of fibrinogens range from 320 to 400 kD, depending on the species of origin, almost all the variability being due to differences in

*Russell F. Doolittle—Departments of Chemistry and Biochemistry and Molecular Biology, University of California, San Diego, La Jolla, California 92093-0314 USA.
Email: rdoolittle@ucsd.edu

Fibrous Proteins, edited by Thomas Scheibel. ©2008 Landes Bioscience.

Table 1. Some properties of human fibrinogen

Molecular mass	340,000
% α helix	33
Mol. dimensions (approximate)	60 × 450 Å
Diffusion constant, D20,w	2.9 × 10-7 cm^2/sec.
Frictional coefficient (fl/fo)	2.34
Sedimentation coefficient, (S20,w)	7.9
Extinction coefficient	15.1
Partial specific volume	0.715
Isoelectric point	5.5
Calculated net charge at pH 7.3	–20
Percent carbohydrate	3
Subunit formula	$\alpha_2\beta_2\gamma_2$

Adapted from reference 20.

the lengths of the α chains.[20] In the case of human fibrinogen, the molecular weights are 67, 54 and 49 kD for the individual α, β and γ chains, respectively. The β and γ chains contain contributions from covalently attached carbohydrate clusters.[21] The three kinds of chain are homologous (evolved from a common ancestral chain), although the α chain has undergone a large scale domainal swap during its evolution that has left a distinctive extended region over the course of its carboxyl two-thirds, instead of the globular domain that exists in β and γ chains.[8]

Fibrinogen has an extended and sigmoidal structure (Fig. 1). Identical globular macrodomains exist at the two ends of the covalent dimer, being attached to the central regions by three-stranded

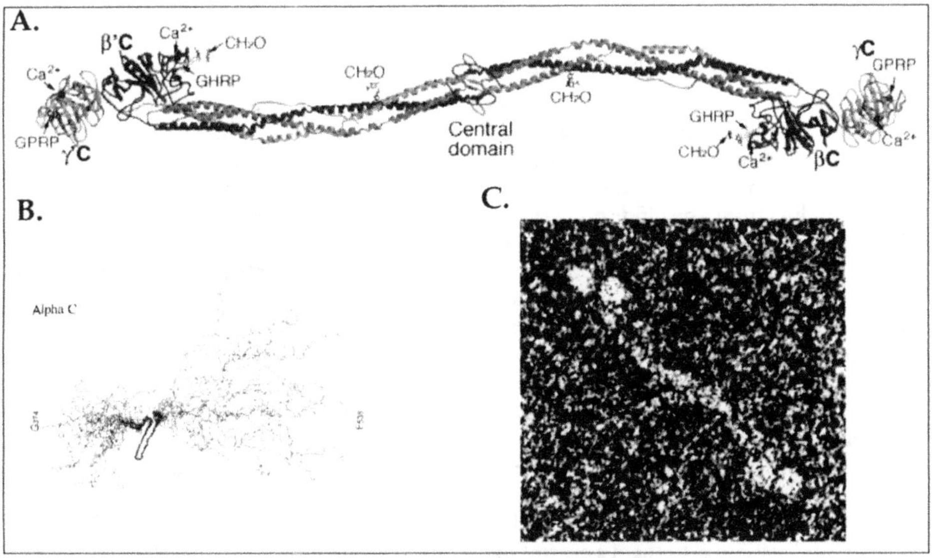

Figure 1. Model of native chicken fibrinogen determined by X-ray crystallography at 2.7 Å (from ref. 16). B) A portion of the αC domain of bovine fibrinogen as depicted by an overlay of 20 lowest energy structures determined by NMR (from ref. 36). The model is shown at the same scale as the model of chicken fibrinogen in Panel A. [The 165 residues account for about half of the αC domain from human fibrinogen.] C) Negative-stained molecule of bovine fibrinogen as viewed by electron microscopy. Photo courtesy of Robley Williams.

coiled coils. The arrangement of chains is such that all six amino termini—including the two pairs of fibrinopeptides—are gathered together in the central part of the molecule around a twofold axis of symmetry. The coiled-coil connectors are vulnerable to proteolysis and cleavages in these regions, as well as elsewhere, give rise to a set of core fragments, the most important of which are called fragments E and D.[22] Fragment E roughly encompasses the central domain and the two fragments D correspond to the terminal regions. Together these two kinds of fragment account for almost two-thirds of the mass of a fibrinogen molecule. Much of the missing third is encompassed by a readily proteolyzed and elusive region termed the αC domain, a structure which in mammals, is attached to the main body of the protein by sets of 13- or 18-residue repeats, the numbers of each differing from species to species.[23]

Numerous posttranslational events occur that affect fibrinogen structure,[24] several of these being species-dependent, also. For example, in many vertebrates, but not higher primates, the fibrinopeptides B contain a sulfated tyrosine. In higher primates, but not in many other mammals, the fibrinopeptide A contains a phosphorylated serine. These particular modifications are in line with most fibrinopeptides having a distinct electronegative charge. Cyclized glutamines at the amino-termini of α and β chains in various species also contribute to this effect. Elsewhere in the molecule, the αC domain in human fibrinogen contains phosphoserine.[24] In almost all species examined, there are carbohydrate clusters on the β and γ chains.

Minor forms of fibrinogen exist that are the result of alternative splicing. In the case of γ chains, somewhere between 5 and 15% of them run through a splice site during translation to yield a longer form called γ'.[25,26] The novel carboxyl-terminal segment resulting from this splice contains a sulfated tyrosine.[27] Importantly, this region has been shown to bind thrombin and factor XIII.[28,29]

A small fraction of unique fibrinogen molecules is the result of another splicing alternative that occurs near the end of the α chain. In this case an extended α chain is generated with a globular domain that is homologous to the globular carboxyl domains of β and γ chains.[30] The function of these extended forms of fibrinogen remains unknown, although they have persisted throughout the course of vertebrate evolution and must be important.

Flexible Portions of Fibrinogen

Even though a crystal structure for a native fibrinogen is in hand,[16] not every architectural feature of the molecule has been revealed. Indeed, several very important regions have proved difficult to pin down. These include the fibrinopeptides, both sets of which are located at the ends of flexible tethers. In the case of the fibrinopeptide A, its thrombin cleavage point is only a dozen residues from a disulfide bond that joins the two halves of the dimeric parent molecule, limiting its sphere of influence. In the case of fibrinopeptide B, however, the thrombin cleavage point is 50 residues away from another disulfide spanning the central axis, leaving the whereabouts of the fibrinopeptide B quite uncertain. The carboxyl terminal segments of γ chains, which contain key cross-linking sites,[31] are also mobile and have not been completely resolved in various electron density maps.

But it is the carboxyl terminal segments of α chains that are the most elusive and that have sparked the most discussion over the years. These regions are readily cleaved from fibrinogen by a variety of proteases and have often been described as "free swimming appendages".[20] In the eyes of some, these parts of the molecule have all the characteristics of what has come to be called "natively unfolded."[32]

There is another school of thought, however, which regards these entities as "compact globules" and attributes to them a central role in fibrin formation.[33-35] Recent NMR studies[36,37] have gone a long way towards resolving the dispute and it may be that both points of view will survive: to wit, in bovine fibrinogen there appears to be a 35-residue ordered compact structure within the mostly unfolded 220-residue αC domain (Fig. 1B).

Early Steps in Fibrin Formation

The finding that two different kinds of peptide are released by thrombin from two different polypeptide chains led to the prescient suggestion half a century ago that the release of one peptide, the fibrinopeptide A, allows the fibrin units to polymerize in a linear mode, whereas the release of the other peptide, the fibrinopeptide B, is responsible for the lateral association of the fibers.[38] The suggestion has been borne out by numerous experiments in the time since, beginning with the finding that the fibrinopeptide A is released at a much faster rate than the B.[39] Today there is general agreement that the bulk of the fibrinopeptide B is not released until the protofibril stage and that the relative rate of its release can influence the character of the final fibrin clot.[40] The differing properties of a clot can include its rigidity, porosity, overall strength and ease of destruction by fibrinolytic enzymes. Morphologically, the fibers may vary in thickness and degree of branching. Arguably, the main cause of these differences will be the immediate concentration of thrombin.

It should also be mentioned that it was long ago proposed that there is a correlation between fibrinopeptide release and the formation of two different kinds of cross-links.[41] Follow-up experiments showed that cross-links between γ chains are introduced before gelation and likely at the protofibril stage,[20] but those involving α chains, which occur at a much reduced rate,[42] depend on the lateral association of the protofibrils.[20]

The removal of a fibrinopeptide—whether it be an A or a B—has two immediate consequences. First, a significant amount of negative charge is removed from the parent molecule, thereby facilitating the mutual approach of these negatively charged entities. Second, a positively charged polymerization site is exposed. The discovery that small synthetic peptides patterned on the newly exposed sequences could prevent the polymerization of fibrin led to the general concept that the positively charged sites were "knobs" that fitted into negatively charged "holes" on the termini of other molecules.[43,44] (The logic was that it is simpler to make a knob than a hole; certainly it was not possible to conceive of a synthetic hole composed of only three or four amino acids). The knob-hole natüre of fibrin formation was completely borne out by X-ray crystallography of fibrinogen and fragments D and D-dimer with synthetic knobs bound in their holes (Figs. 2 and 3). More recently, evidence has been reported that implicates regions of the uncleaved fibrinopeptide B in protofibril assembly.[45,46]

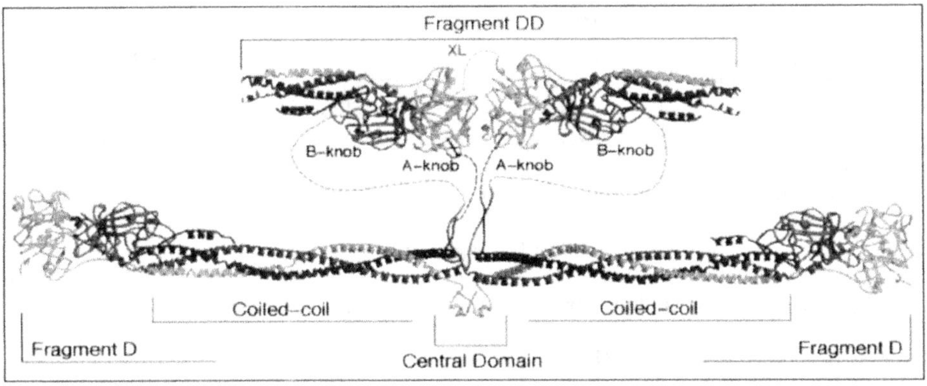

Figure 2. Proposed knob-hole (kh) interactions between central domain and γC (green) and βC (blue) domains of two end-to-end associated molecules within a single protofibril. The 80 Å-distance between the two strands of the protofibril is an approximation based on the extrapolation of residues γ34 back to γ17 (red dashed line), which are not clearly delineated in the chicken fibrinogen structure. In this depiction, the longer flexible regions of β chains (residues β64 back to β15, blue dashed lines) allow the B knobs to bind to the same two companion molecules in a clasper-style (from ref. 54). A color version of this figure is available at www.eurekah.com.

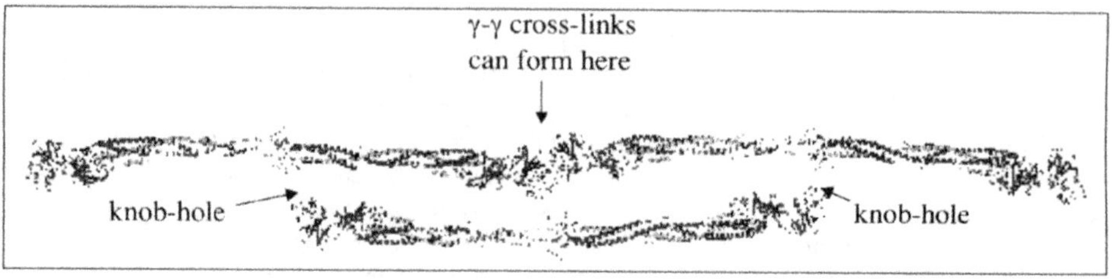

Figure 3. Model of three fibrin(ogen) molecules as they would occur at very earliest stage of clotting. The central region of one molecule is associated with terminal regions of two other molecules, supplying the pair of A knobs that bind to the γC holes of the others.

The Association of Protofibrils

The question arises, how do protofibrils interact with each other to form fibers? This remains a very active area of inquiry and there is not universal agreement on how the process occurs. It is also the stage at which the defining features of differing forms of fibrin are likely introduced. Just as there have been differing views about the structure of αC domains, so has there been a corollary disagreement about their role in the association process.

In this regard, in 1985, a very appealing hypothesis was put forward regarding the role of the αC domain in polymerization.[33] Building on observations made by electron microscopy that the central region of fibrinogen contained more mass than could be accounted for by fragment E[47,5] and also that the terminal parts of the α chains were connected to the main body of the parent molecule by flexible segments with random coil propensities,[8] the authors postulated that the two αC domains in a fibrinogen molecule were associated intra-molecularly, but during conversion to fibrin they became associated *inter*-molecularly. The proposal came to be known as the "intra- to inter- switch hypothesis."[48-50]

Numerous studies have been conducted in the interval in attempts to validate the hypothesis. Especially, various kinds of electron microscopy were employed and many elegant and convincing micrographs were published that seemed to support the idea. During the course of these studies, it was further proposed that it was the release of the fibrinopeptide B that allowed the rearrangement of αC domains to occur (Fig. 4).[48,49]

As it happened, many of the earlier experiments had utilized partially proteolyzed fibrinogens lacking one or both αC domains, even though such preparations almost always lack the amino-terminal segments of β chains, including the now implicated fibrinopeptide B. Later studies used recombinant fibrinogens that were specificially engineered to lack the αC domain only; other experiments were conducted that used recombinant constructs of all or parts of the αC domains themselves. Not all of the studies with recombinant material were in full agreement with earlier findings and the endorsement of the hypothesis is not quite as resounding as it seemed a decade ago.[51] Still, it should be noted that recent NMR experiments on shortened forms of recombinant αC domains have shown that the intrinsic instability of these domains can be significantly reduced by cooperative interactions of the sort expected as a result of intermolecular association.[37]

A Somewhat Different View

A contrary view is based partly on skepticism about the specificity of the alleged interactions. As noted above, αC domains vary greatly from species to species, making it difficult to envision a specific set of interactions. It is indisputable that both the fibrinopeptides B and the αC domains change extremely rapidly during evolution and as such would have difficulty maintaining specific interactions.

Not coincidentally, neither of these regions of the fibrinogen molecule has been visualized in electron density maps calculated from X-ray diffraction of crystals of native fibrinogen, indicating that they are extremely mobile.[16] Clearly, whatever interactions they might enjoy in solution

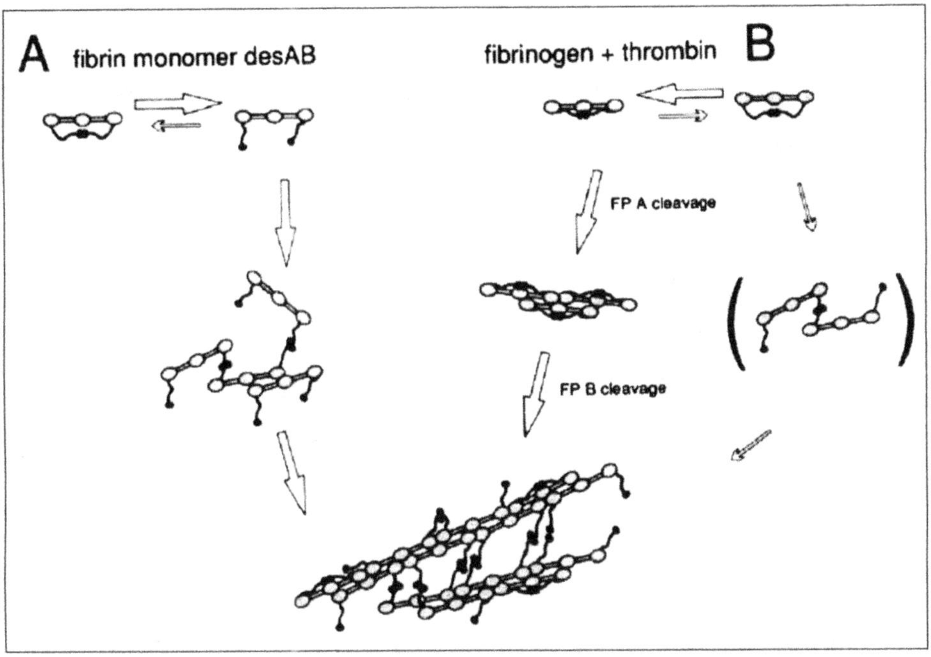

Figure 4. Schematic depiction of the "intra- to inter- switch hypothesis." It has been proposed that initially αC domains are associated *intra*molecularly in fibrinogen, but they become associated *inter*molecularly as a result of the release of fibrinopeptide B (from ref. 49).

do not extend to any kind of contacts that would hold them in place in a crystal. Nor has it been possible to crystallize these domains from any species, whether prepared from native fibrinogen or expressed in recombinant forms.[32]

The susceptibility of the αC domain to proteolysis is another symptom of its mostly unfolded structure. There is also the matter of α chain cross-linking and what it implies about associations between αC domains in fibrin: namely, there are many different cross-link donors and acceptors,[52,53] implying that linked αC domains are the product of random encounters. Finally, the recently reported NMR structures[37,38] reveal the open and mostly extended nature of these domains, a compact core notwithstanding (Fig. 1B).

An alternative model for how protofibrils associate is based on interactions observed in crystals of core fragments of fibrinogen and fibrin in the presence and absence of synthetic peptide knobs.[54] (Figs. 5-8). Specifically, when the γC hole is occupied by the A-knob, the molecules tend to associate, first across a specific end-to-end γ-γ interface and then, at the protofibril stage, across a second side-by-side γ-γ interface (Fig. 5). The second step would be facilitated merely by the loss of the highly negatively charged fibrinopeptide B.

In this model, the interactions of αC domains are more a consequence of lateral association than a cause. The principal virtue of this model is that it describes specific interactions between specified parts of each fibrin unit, thereby keeping the assembled protofibrils in register, as is required by the fact that fibrin strands maintain a 230 Å periodicity. Moreover, the calculated density of a model fibrin strand is in good agreement with experimentally determined values.[54]

The radial nature of the growing fiber in the model is consistent with another interesting feature of fibrin formation: namely, most fibrin fibers under the same conditions tend to be about the same width, some inherent feature of their formation being self-limiting.[55] It has been proposed that the self-limiting nature of lateral growth is related to the fibers being observably twisted, the twist of necessity stretching the outermost protofibils and forcing a natural limit.[55]

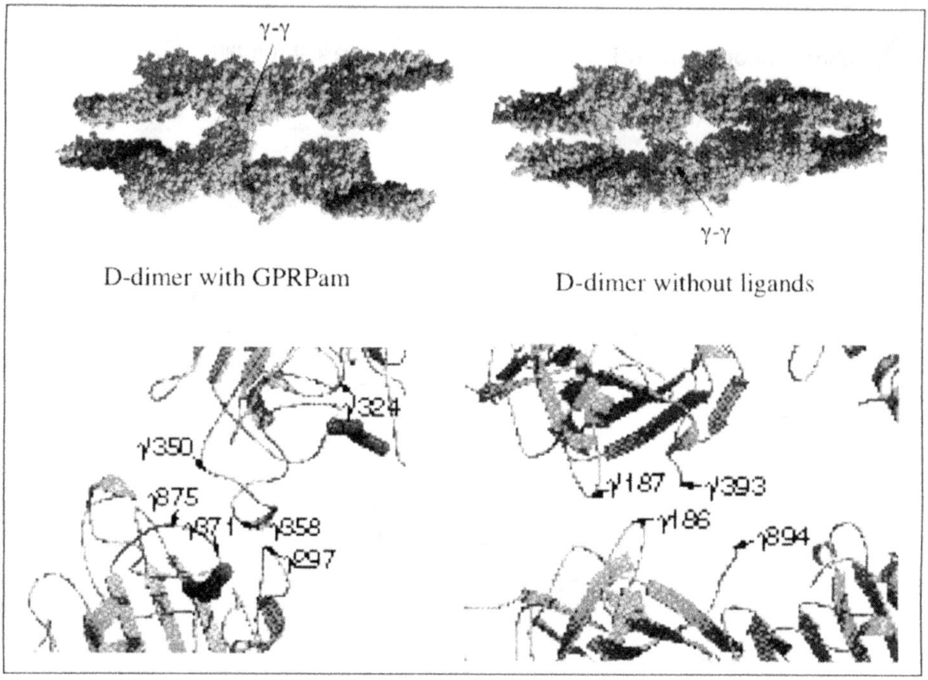

Figure 5. Comparison of crystal contacts between D-dimers that were crystallized in presence and absence of synthetic A knobs. The circled regions in the top panel show the very different "side-by-side" associations of γC units that occur. The bottom panel shows a close-up of the circled region and includes residue numbers. Note that the residues involved in the presence of the A knob corresponds to the peptide reported by Demfle et al[56] as greatly influencing fibrin formation. Green, γ chains; blue, β chains, red, α chains. (From ref. 54.) A color version of this figure is available at www.eurekah.com.

Recent support for this model emerged from an unexpected source. In a Herculean effort, Dempfle et al synthesized 76 overlapping dodecameric peptides corresponding to the full length of the γC domain.[56] Three of these were found to influence fibrin formation greatly, especially one covering the stretch encompassed by residues γ369-380. Another peptide covering the stretch γ350-362 also had a significant effect, as did one spanning the region γ330-342. Coincidentally, two of the three regions, including the most effective one, match the interaction sites specified in the model (Fig. 5).

Resolving the Differences

How are we to decide between these two different views of how protofibrils associate? As often happens, both views may have some redeeming features and both kinds of participation may occur. Certainly the evidence favoring a role for αC domains in fibrin formation continues to accumulate[57-59] and, most recently, experiments with laser optical tweezers have shown an involvement of the central domain of fibrinogen with the αC domain.[60] These experiments were conducted with fragments from the central domain lacking one or the other of the fibrinopeptides (or neither or both) binding to a recombinant fibrinogen with defective γC holes. Significant interactions were detected, not only between entities containing fibrinopeptides B, on the one hand and αC domains, on the other, but also—surprisingly—between moieties with fibrinopeptides A on one side and fragments D or E on the other.

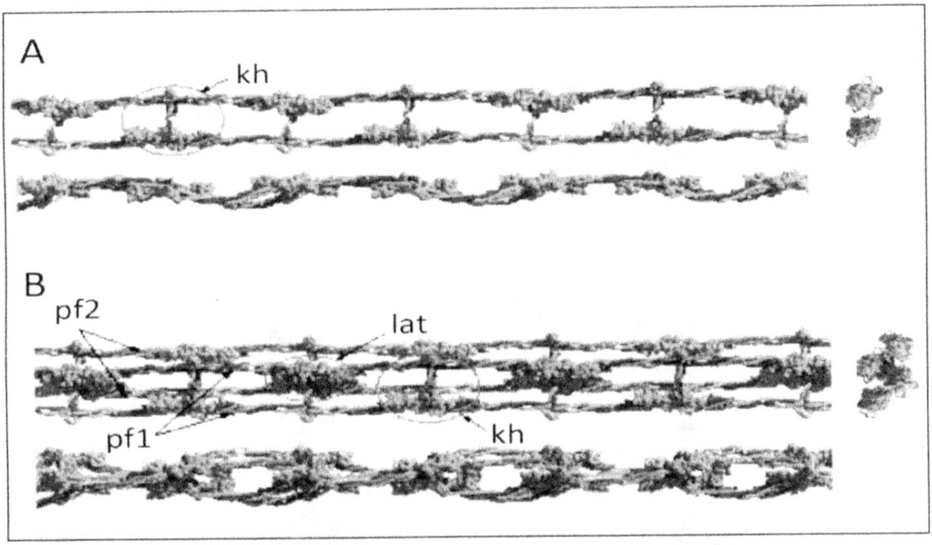

Figure 6. A) Three views (top, side and end-on) of a model protofibril formed by knob-hole (kh) interactions. The protofibrils, which may be 10-20 units in length, have a ribbon-like structure. B) Same three views of the lateral (lat) association of two protofibrils (pf1 and pf2). Color scheme as in Figure 5. (From ref. 54.)

In any case, all the interactions reported are weak and transient, not firm and specific attachments. The fact that fibrinopeptides A may be involved in these interactions, as well as the anticipated fibrinopeptides B, further suggests that what is being observed is a general electrostatic effect.

Nonetheless, these are likely real effects and although they may not result in enduring interactions, they doubtless influence fibrin formation. Others have suggested in the past that the role of αC domains is mainly a guiding or accelerating one,[33,49] a quite reasonable suggestion, whatever the basis of the guidance and acceleration may be.

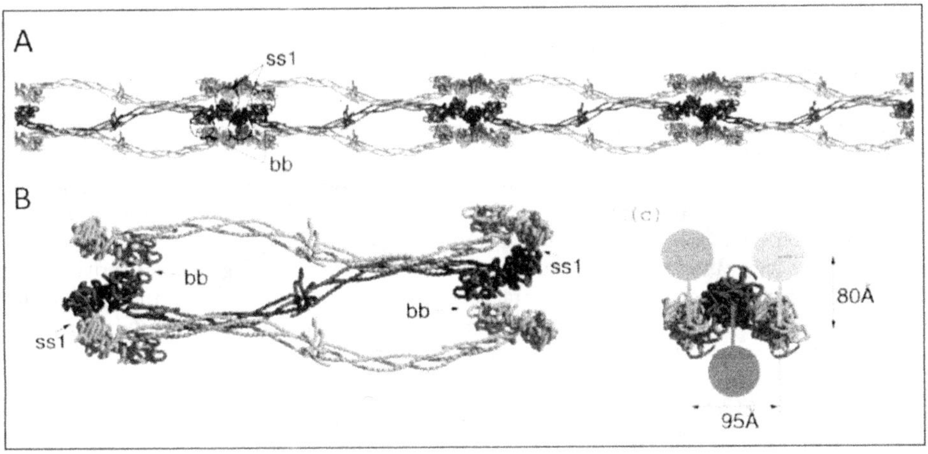

Figure 7. Three protofibrils that would result from the lateral association of γC domains. For purposes of visualization, only one of the strands from each protofibril is shown; the second set is shown as solid circles in the end-on view. Color scheme as in previous figures. (From ref. 54.)

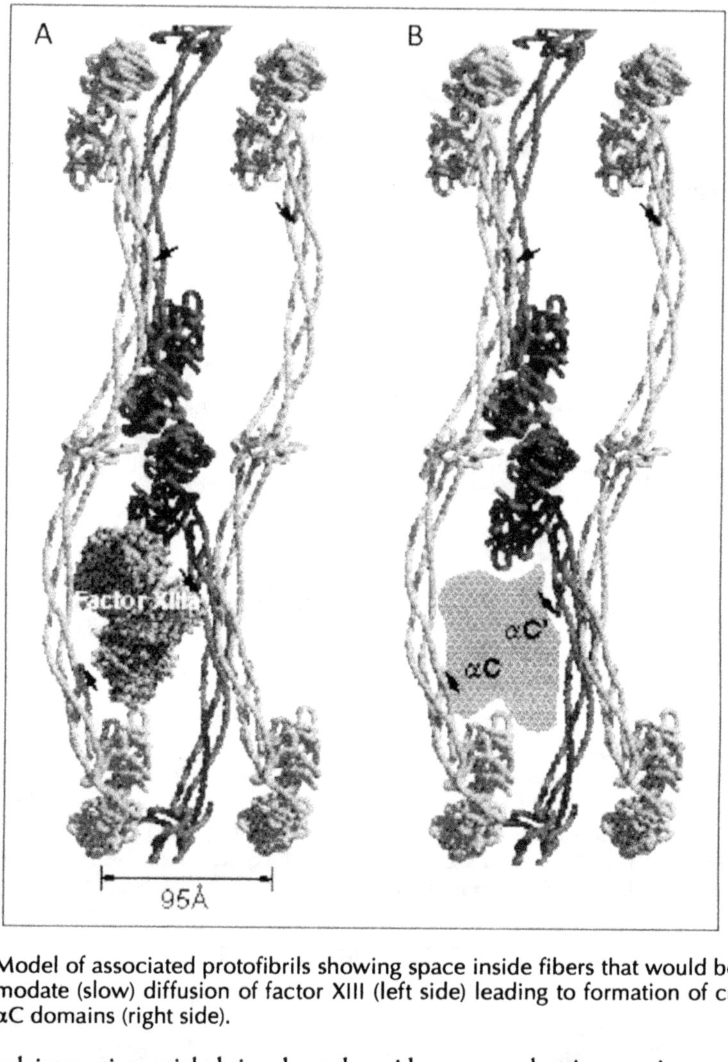

Figure 8. Model of associated protofibrils showing space inside fibers that would be needed to accommodate (slow) diffusion of factor XIII (left side) leading to formation of cross-links between αC domains (right side).

How such interactions might bring about the guidance or acceleration remains mysterious. It may be, for example, that these highly mobile and polar regions are serving as fenders that actually prevent premature association of protofibrils. *A priori*, the association of protofibrils ought to be most effective at some optimum degree of overlap. For example, if two protofibrils overlap one hundred percent, the fiber won't grow linearly. If the degree of overlap is too little, the associative forces will be very weak. Finding the optimum overlap where the fiber will grow to maturity in both the linear and lateral modes may take several collisions. It is not impossible that the αC domains serve as chaperones, either preventing some undesirable collisions or allowing second chances for others. Future experiments will doubtless settle the matter.

Acknowledgements

This review was written during the tenure of NIH Grant HL-81,553.

References

1. Ferry JD. The mechanism of polymerization of fibrin. Proc Natl Acad Sci, USA 1952; 38:566-569.
2. Matacic S, Loewy AG. Transglutaminase activity of the fibrin cross-linking enzyme. Biochem Biophys Res Comm 1966; 24:858-866.
3. Hall CE, Slayter HS. The fibrinogen molecule: its size, shape and mode of polymerization. J Biophys Biochem Cytol 1959; 5:11-16.
4. Williams RC. Morphology of bovine fibrinogen monomers and fibrin polymers. J Mol Biol 1981; 150:399-408.
5. Erickson HP, Fowler WE. Electron microscopy of fibrinogen, its plasmic fragments and small polymers. Ann NY Acad Sci 1983; 408:146-163.
6. Lottspeich F, Henschen A. Amino acid sequence of human fibrin. Preliminary note on the γ-chain sequence. Hoppe-Seyler's Z Physiol Chem 1977; 358:935-938.
7. Henschen A, Lottspeich F. Amino acid sequence of human fibrin. Preliminary note on the completion of the β-chain sequence. Hoppe-Seyler's Z Physiol Chem 1977; 358:1643-1646.
8. Doolittle RF, Watt KWK, Cottrell BA et al. The amino acid sequence of the α-chain of human fibrinogen. Nature 1979; 280:464-468.
9. Yee VC, Pratt KP, Cote HC et al. Crystal structure of a 30 kDa C-terminal fragment from the γ chain of human fibrinogen. Structure 1997; 5:125-138.
10. Pratt KP, Cote HCF, Chung DW et al. The fibrin polymerization pocket: three dimensional structure of a 30-kDA C-terminal γ chain fragment complexed with the peptide Gly-Pro-Arg-Pro. Proc Natl Acad Sci USA 1997; 94:7176-7181.
11. Spraggon G, Everse SJ, Doolittle RF. Crystal structures of fragment D from human fibrinogen and its crosslinked counterpart from fibrin. Nature 1997; 389:455-462.
12. Everse SJ, Spraggon G, Veerapandian L et al. Crystal structure of fragment double-D from human fibrin with two different bound ligands. Biochemistry 1998; 37:8637-8642.
13. Spraggon G, Applegate D, Everse SJ et al. Crystal structure of a recombinant αEC domain from fibrinogen-420. Proc Natl Acad Sci USA 1998; 95:9099-9104.
14. Everse SJ, Spraggon G, Veerapandian L et al. Conformational changes in fragments D and double-D from human fibrin(ogen) upon binding the peptide ligand Gly-His-Arg-Pro-amide. Biochemistry 1999; 38:2941-2946.
15. Brown JH, Volkmann N, Jun G et al. The crystal structure of modified bovine fibrinogen. Proc Natl Acad Sci USA 2000; 97:85-90.
16. Yang Z, Kollman JM, Pandi L et al. Crystal structure of a native chicken fibrinogen at 2.7 Å resolution. Biochemistry 2001; 40:12515-12523.
17. Modrazo J, Brown JH, Litvinovich S et al. Crystal structure of the central region of bovine fibrinogen (E5 fragment) at 1.4 Å resolution. Proc Natl Acad Sci USA 2001; 98:11967-11972.
18. Blomback B, Yamashina I. On the N-terminal amino acids in fibrinogen and fibrin. Arkiv Kemi 1958; 12:299-319.
19. Henschen A. Number and reactivity of disulfide bonds in fibrinogen and fibrin. Arkiv Kemi 1964; 22:355-374.
20. Doolittle RF. Structural aspects of the fibrinogen to fibrin conversion. Adv Prot Chem 1973; 27:1-105.
21. Townsend RR, Hilliker E, Li Y-T et al. Carbohydrate structure of human fibrinogen. J Biol Chem 1982; 257:9704-9710.
22. Nussenzweig V, Seligmann M, Pelimont J et al. Les produits de degradation du fibrinogene humain par le plasmine. Ann Inst Pasteur 1961; 100:377-389.
23. Murakawa M, Okamura T, Kamura T et al. Diversity of primary structures of the carboxy-terminal regions of mammalian fibrinogen Aα chains. Thromb Haemostasis 1993; 69:351-360.
24. Henschen-Edman AH. Fibrinogen non-inherited heterogeneity and its relationship to function in health and disease. Ann NY Acad Sci 2001; 936:580-593.
25. Fornace AJ, Cummings DE, Comeau CM et al. Structure of human γ-fibrinogen gene. Alternate nRNA splicing near the 3' end of the gene products γA and γB forms of γ-fibrinogen. J Biol Chem 1984; 259:12826-12830.
26. Chung DW, Davie EW. γ and γ' chains of human fibrinogen are produced by alternative mRNA processing. Biochemistry 1984; 23:4232-4236.
27. Hirose S, Oda K, Ikehara Y. Tyrosine O-sulfation of the fibrinogen γB chain in primary cultures of rat hepatocytes. J Biol Chem 1988; 263:7426-7430.
28. Siebenlist KR, Meh DA, Mosesson MW. Plasma factor XIII binds specifically to fibrinogen molecules containing γ' chains. Biochemistry 1996; 35:10448-10453.
29. Meh DA, Siebenlist KR, Brennan SO et al. The amino acid sequence in fibrin responsible for high affinity thrombin binding. Thromb Haemostst 2001; 85:470-474.

30. Fu Y, Cao Y, Hertzberg KM et al. Fibrinogen α genes: conservation of bipartite transcripts and carboxy-terminal extended α subunits in vertebrates. Genomics 1995; 30:71-76.
31. Chen R, Doolittle RF. γ-γ Cross-linking sites in human and bovine fibrin. Biochemistry 1971; 10:4486-4491.
32. Doolittle RF, Kollman JM. Natively unfolded regions of the vertebrate fibrinogen molecule. Proteins 2006; 63:391-397.
33. Medved LV, Gorkun OV, Manyakov VF et al. The role of fibrinogen alpha C-domains in the fibrin assembly process. FEBS Letts 1985; 181:109-112.
34. Weisel JW, Stauffacher CV, Bullitt E et al. A model for fibrinogen: domains and sequence. Science 1985; 230:1388-1391.
35. Medved L, Weisel JW. The structure and function of the αC domains of fibrinogen. Ann NY Acad Sci 2001; 936:312-327.
36. Burton RA, Tsurupa G, Medved L et al. Identification of an ordered compact structure within the recombinant bovine fibrinogen αC-domain fragment by NMR. Biochemistry 2006; 45:2257-2266.
37. Burton RA, Tsurupa G, Hantgan RR et al. NMR solution structure, stability and interaction of the recombinant bovine fibrinogen αC-domain fragment. Biochemistry 2007; 46:8550-8560.
38. Laurent T, Blomback B. On the significance of the release of two different peptides from fibrinogen during clotting. Acta Chem Scand 1958; 12:1875-1878.
39. Shainoff JR, Page IH. Cofibrins and fibrin intermediates as indicators of thrombin acitivity in vivo. Circ Res 1960; 8:1013-1022.
40. Weisel JW, Veklich Y, Gorkun O. The sequence of cleavage of fibrinopeptides from fibrinogen is important for protofibril formation and enhancement of lateral aggregation in fibrin clots. J Mol Biol 1993; 232:285-297.
41. Doolittle RF, Cassman KG, Chen R et al. Correlation of the mode of fibrin polymerization with the pattern of cross-linking. Ann N Y Acad Sci 1972; 202:114-126.
42. McKee PA, Mattock P, Hill RL. Subunit structure of human fibrinogen, soluble fibrin and cross-linked insoluble fibrin. Proc Natl Acad Sci USA 1970; 66:738-744.
43. Laudano AP, Doolittle RF. Synthetic peptides which bind to fibrinogen and prevent polymerization of fibrin monomers. Proc Natl Acad Sci USA 1978; 75:3085-3089.
44. Laudano AP, Doolittle RF. Studies on synthetic peptides that bind to fibrinogen and prevent fibrin polymerization. Structural requirements, number of binding sites and species differences. Biochemistry 1980; 19:1013-1019.
45. Pechik I, Yakovlev S, Mosesson MW et al. Structural basis for sequential cleavage of fibrinopeptides upon fibrin assembly. Biochemistry 2006; 45:3588-3597.
46. Lugovskoy EV, Gritsenko PG, Kapustianenko IN et al. Functional role of Bβ-chain N-terminal fragment in the fibrin polymerization process. FEBS J 2007; 274:4540-4549.
47. Wall J, Hainfield J, Haschemeyer RH, Mosesson MW. Analysis of human fibrinogen by scanning transmission electron microscopy. Ann NY Acad Sci 1983; 408:164-179.
48. Veklick YI, Gorkun O, Medved L et al. Carboxyl-terminal portions of the alpha chains of fibrinogen and fibrin. Localization by electron microscopy and the effects of isolated alpha C fragments on polymerization. J Biol Chem 1993; 268:13577-13585.
49. Gorkun OV, Veklich YI, Medved LV et al. Role of the αC domains in fibrin in clot formation. Biochemistry 1994; 33:6986-6997.
50. Tsurupa G, Veklich Y, Hantgan R et al. Do the isolated fibrinogen αC-domains form ordered oligomers? Biophys Chem 2004; 112:257-266.
51. Gorkun OV, Henschen-Edman AH, Ping LF et al. Analysis of Aα251 Fibrinogen: The αC domain has a role in polymerization, albeit more subtle than anticipated from the analogous proteolytic fragment X. Biochemistry 1998; 37:15434-15441.
52. Sobel JH, Gawinowicz MA. Identification of α chain lysine donor sites involved in factor XIIIa fibrin cross-linking. J Biol Chem 1996; 271:19288-19297.
53. Matsuka YV, Medved LV, Migliorini MM et al. Factor XIIIa-catalyzed cross-linking of the recombinant αC fragments of human fibrinogen. Biochemistry 1996; 35:5810-5816.
54. Yang Z, Mochalkin I, Doolittle RF. A model of fibrin formation based on crystal structures of fibrinogen and fibrin fragments complexed with synthetic peptides. Proc Natl Acad Sci USA 2000; 97:14156-14161.
55. Weisel JW, Nagaswami C, Makowski L. Twisting of fibrin fibers limits their radial growth. Proc Natl Acad Sci USA 1987; 84:8991-8995.
56. Dempfle C-E, Bindeballe N, Munchback M et al. Changes in fibrinogen and fibrin induced by a peptide analog of fibrinogen γ365-380. J Thromb Haemost 2007; 5:1707-1714.
57. Litvinov RJ, Gorkun OV, Owen S et al. Polymerization of fibrin: specificity, strength and stability of knob-hole interactions studied at the single-molecule level. Blood 2005; 106:2944-2951.

58. Collet J-P, Moen JL, Veklich YI et al. The αC domains of fibrinogen affect the structure of the fibrin clot, its physical properties and its susceptibility to fibrinolysis. Blood 2005; 106:3824-3830.
59. Gorkun OV, Litvinov RI, Veklich YI et al. Interaction mediated by the N-terminus of fibrinogen's Bβ Chain. Biochemistry 2006; 45:14843-14852.
60. Litvinov RI, Yakovlev S, Tsurpa G et al. Direct evidence for specific interactions of the fibrinogen αC-domains with the central E region and with each other. Biochemistry 2007; 46:9133-9142.

CHAPTER 5

Structure, Function and Assembly of Flagellar Axial Proteins

Ferenc Vonderviszt* and Keiichi Namba

Abstract

The bacterial flagellum is a biological macromolecular nanomachine for locomotion. A membrane embedded molecular motor rotates a long helical filament that works as a propeller driving the bacterium through the liquid environment. The flagellum is composed of about 30 different proteins with copy numbers ranging from a few to a few thousands and is made by self-assembly of those proteins. The helical filament can be transformed into various distinct supercoiled forms by changes in chemical environment, single amino acid mutations, or mechanical forces. The axial portion of the flagellum involves several substructures: the rod, the hook, the hook-filament junction, the long helical filament and a cap at the filament tip. Although the axial component proteins are dissimilar at the primary sequence level, they share common structural characteristics. The central portions of their amino acid sequences are hypervariable whereas the amino- and carboxy-terminal parts contain highly conserved segments. In their monomeric form, they posses large, natively-disordered terminal regions that are essential in controlling and mediating intersubunit interactions. Recent structural studies—combining X-ray diffraction and single particle image analysis by cryo-EM—have given insights into the molecular mechanisms of self-assembly, supercoiling and polymorphic ability and open the way for construction of various flagella-based systems for applications in vaccination, bio- or nanotechnology.

Introduction

Flagella are the organelles for bacterial locomotion. These supramolecular structures extend from the cytoplasm to the cell exterior and are composed of three major structural elements, the basal body, the hook and the filament (Fig. 1).[1,2] The swimming pattern of bacteria, such as *Escherichia coli* and *Salmonella*, consists of straight swimming for a few seconds and tumbling for a fraction of a second. During the straight swimming phase, the helical filaments form a bundle behind the cell body, where the filaments are all in a left-handed supercoiled form, each acting as a propeller driven by a rotary motor at the base of the flagellum. Bacteria tumble once every few seconds to change their swimming direction for the chemotactic behavior. The tumbling is triggered by quick reversal of motor rotation, which produces a twist in the filament structure and transforms it into right-handed supercoils momentarily. This allows the bundle to fall apart smoothly and then the uncoordinated propelling forces change the orientation of the cell quickly.[3,4] The flagellar filaments can also be transformed into various but distinct polymorphic forms including two straight forms, in response to amino acid replacements in flagellin[5] and to chemical changes in the environment[6,7] as well as to mechanical forces.[3,8]

The long helical filament is connected to the basal structure via a short, highly curved segment called the hook.[9] Both the hook and the helical filament are self-assembling macromolecular

*Corresponding Author: Ferenc Vonderviszt—Department of Nanotechnology, Faculty of Information Technology, University of Pannonia, Egyetem u. 10, H-8200 Veszprém, Hungary. Email: von007@almos.vein.hu

Fibrous Proteins, edited by Thomas Scheibel. ©2008 Landes Bioscience.

structures composed of the hook protein (FlgE) and flagellin (FliC), respectively. Each filament may comprise as many as ~30000 flagellin subunits and can grow up to ~15 μm. The hook is a helical assembly of ~130 copies of FlgE subunits with a well regulated length of 55 nm ± 6 nm,[10] capable of forming polymorphic supercoil structures.[11] Bacterial flagellar hook acts as a molecular universal joint that can transmit the torque produced by the basal body, a rotary motor, to the flagellar filament.

Other components of the filamentous axial portion of the bacterial flagellum are the five rod proteins (FliE, FlgB, FlgC, FlgF, FlgG)[12] and the hook associated proteins HAP1 (FlgK), HAP2 (FliD) and HAP3 (FlgL).[13,14] HAP1 and HAP3 are junction proteins that connect the hook to the filament. HAP2 forms a capping structure at the distal end of the flagellar filament, which

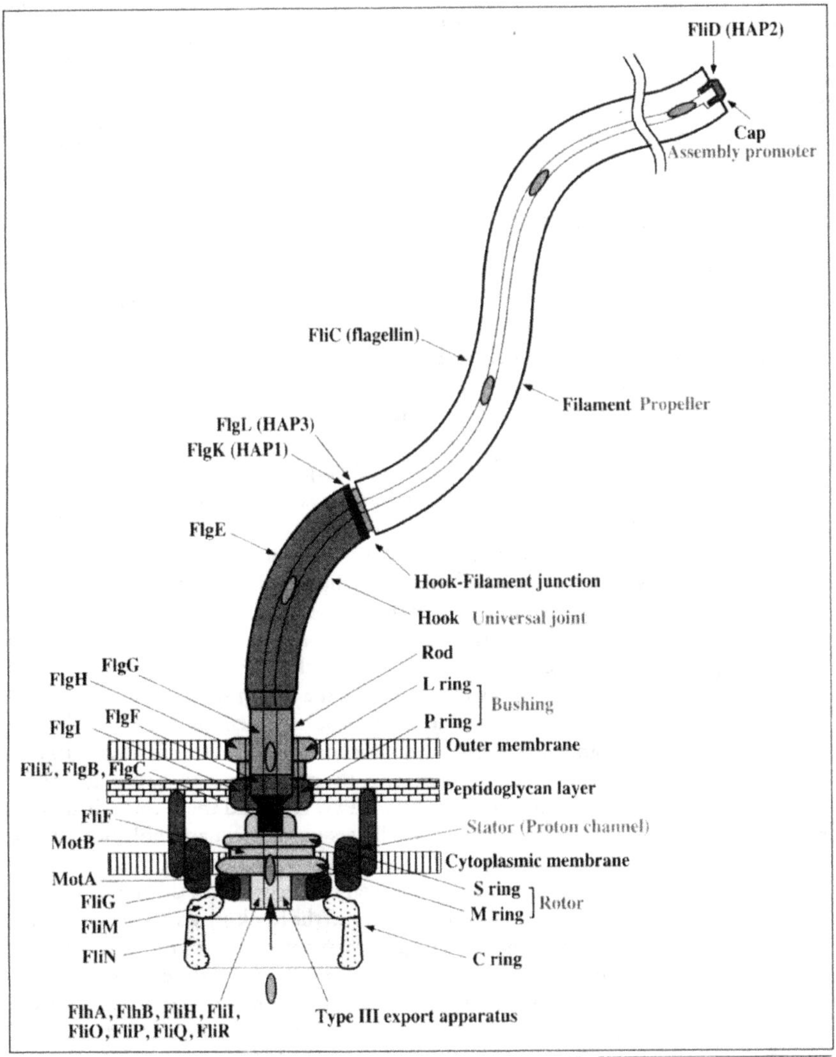

Figure 1. Schematic diagram of the bacterial flagellum. Different colors represent different protein components Reprinted with permission from reference 20.

helps to incorporate flagellin monomers—transported through the central channel—into the filament at the tip.

The flagellar proteins forming the structures lying beyond the cytoplasmic membrane are synthesized in the cell and exported sequentially by the flagellum-specific protein export apparatus from the cytoplasm to the site of assembly at the distal end of the growing filament.[15] They must be translocated through the narrow—20-25 Å wide—central channel of the flagellum in mostly unfolded conformation. The flagellar protein export system is thought to exist at the cytoplasmatic face of the basal body to distinguish flagellar proteins from other cytoplasmatic proteins and to facilitate their transportation. The flagellar protein export system is a specialized type III export machinery.[1,15,16] The type III secretion systems include those for secretion of virulence factors by a wide variety of pathogenic bacteria.

In this review we will focus on the external filamentous portion of the bacterial flagellum, the structural characteristics of the axial component proteins, their role in filament assembly and polymorphism and applications in bio- and nanotechnology. Results obtained mainly with *E. coli* and *Salmonella enterica* serovar *Typhimurium* will be described. The flagellar motor, the flagellum-specific export machinery and several other interesting topics related to the flagellar system lie beyond the scope of this chapter. There are excellent recent reviews discussing these topics.[15,17-19]

Flagellar Filament Assembly

In Vivo Assembly

The assembly process of the bacterial flagellum starts from the formation of the FliF ring complex (also called the MS ring) of the basal body in the cytoplasmic membrane and proceeds in both inward and outward directions, as well as laterally.[1,20,21] The inward assembly involves the formation of the C ring, which is also called the switch complex, in the cytoplasmic space and the flagellar export apparatus is formed within the C ring, getting ready to export flagellar axial proteins through its central channel.

For all axial structure formation, the flagellum-specific type III protein export system selectively binds and translocates flagellar axial proteins into the central channel of the flagellum.[15] Since the identification and enzymatic characterization of FliI ATPase as a component of the flagellar protein export system,[22] it had been thought that the flagellar protein export is driven by the energy of ATP hydrolysis. A recent study, however, clearly showed that the flagellar proteins are exported to form the flagellum even in the absence of FliI and that the proton motive force across the cytoplasmic membrane is responsible for driving most part of the export process that involves unfolding of export substrate proteins and translocation of the unfolded chains, with a help of the FliI hexamer ring complex for insertion of the NH_2-terminal chain of the export substrate proteins.[83]

The axial proteins construct the rod, the hook, the hook-filament junction and the long filament in this order. The rod composed of five proteins is connected to the FliF ring at its proximal portion and with the hook at its distal end and traverses the periplasmic space through the LP ring, where the P ring is located within the peptidoglycan (PG) layer and the L ring within the outer membrane. The rod cap made of FlgJ not only facilitates rod protein assembly but also makes a local hole through the PG layer by its muramidase activity, thus permitting penetration of the rod through the PG layer.[23] Then FlgE assembles in a helical manner to form the hook. The hook cap made of FlgD is attached at the distal end of the hook until the hook grows up to a length of about 55 nm.[24] The FliK and FlhB proteins function together to control the hook length within 10% of the average length.[25-27] The hook length is determined at the point when the export system switches its export specificity from recognition of early substrates called rod/hook type (rod proteins, FlgJ, hook protein, FlgD, FliK) to late substrates called filament type (FlgM, HAP proteins, flagellin).[28] Then the FlgD cap falls off the tip of the hook and is replaced by HAP1 and then HAP3 and HAP2 are bound at the distal end in this order to form the basal body-hook-HAP1-HAP3-HAP2 complex momentarily before the initiation of flagellin assembly into the long helical filament.[13,14] Flagellin monomers travel a long way through the narrow central channel,[29] only 2 nm in diameter but up to 10-15 μm long,[30] to the flagellar tip, presumably by a diffusion process.

Filament assembly in vivo requires the presence of the HAP2 pentamer complex capping the distal end of the filaments where flagellin monomers are assembled.[14,29] The cap permits assembly of flagellin and prevents its excretion. The simplest role of the cap is to prevent subunits from diffusing away from the tip but it may also induce a conformational change in flagellin required for polymerization. Interactions between the cap and distal end of the filament seem to have unique characteristics. The binding is supposed to be very stable so that dissociation of the cap rarely occurs and yet flagellin monomers can be easily inserted between the HAP2 cap and the end of the filament.

In Vitro Filament Reconstruction

The flagellar filaments can be easily detached from cells and depolymerized into monomeric flagellin by either heat treatment (60°C for 10 minutes) or acid treatment (around pH 2.5). The flagellin monomers thus obtained are capable of self-assembly. They can be reconstituted into the filaments under physiological conditions by adding short filaments as seeds or a precipitant such as ammonium sulfate.[32] The hook can also be reconstituted from monomeric hook proteins under appropriate conditions in vitro.[33] Moreover, it has been demonstrated that the hook-filament complex can also be formed by sequential addition of HAP1, HAP3, flagellin and HAP2 to the hook.[34]

Overall Structural Characteristics of Axial Proteins

Sequential Similarities

Flagellar axial proteins can be divided into three major families.[35,36] Hook protein shows sequential similarities, especially in its terminal regions, to the rod proteins (FliE, FlgB, FlgC, FlgF, FlgG), which compose the drive shaft of the flagellar motor, and also to the proximal hook associated protein, HAP1. Flagellin is not similar to the hook family, but appears to be related to HAP3. HAP2 is the most dissimilar among the axial proteins; it shows no significant sequential homology to flagellin or hook protein.

Conserved Terminal Regions and Variable Central Portions

In spite of the sequential dissimilarities, comparison of amino acid sequences from different bacterial sources reveals common motifs for flagellar axial proteins. Their terminal regions are highly conserved whereas the central portion contains highly variable segments. For example, flagellins have widely differing molecular masses in the range from 27 kDa to 67 kDa.[37,38] The difference is in the central part of the molecule demonstrating that large deletions and insertions can be introduced into the central region of flagellin without destroying its ability to form the flagellar filaments. The variable central part is known to be the major determinant for flagellar antigenicity.

Heptad Repeats of Hydrophobic Amino Acids in the Terminal Regions

The NH_2- and COOH-terminal regions of the axial components of bacterial flagella possess heptad repeats of hydrophobic amino acid residues,[36] a characteristic of sequences that fold into α-helical coiled coils. It was suggested that the terminal regions of axially neighboring subunits interact to form coiled-coil bundles upon polymerization. This interlocking organization has been postulated to be the common motif by which the axial proteins form a continuous, mechanically stable structure[36] and it has been visualized in the atomic model of one of the two straight-type flagellar filament.[30]

Disordered Terminal Regions and Compact Central Portions

Proteolytic, calorimetric and NMR studies revealed that terminal disorder is a common structural feature of the axial proteins of the bacterial flagellum.[12,39-42] Representative members from all the major groups of the axial protein family have been demonstrated to possess disordered terminal regions. For example, the disordered terminal regions of *Salmonella* flagellin from the SJW1103 strain span the first 66 and the last 44 amino acid residues.[39] The disordered regions belong to the most conserved part of flagellin, indicating their important structural and functional roles. The disordered terminal regions possess a high α-helix forming potential and assemble into helical bundles upon polymerization. The other parts of flagellin form compact, well-folded domains, which involve conserved and variable segments as well.[41,43,44]

The disordered terminal regions possess multiple roles. Polymerization properties of all axial proteins seem to be similarly controlled by their disordered terminal regions.[44-46] Subunits are synthesized in the cytoplasm in large amounts and therefore it is essential to prevent their polymerization within the cell. Monomeric flagellin stays monomeric even at high protein concentrations since subunits with large disordered regions can not effectively bind to each other. The self-assembly of these proteins requires the presence of template structures, therefore polymerization occurs only at the distal end of the growing filament.[29]

The disordered terminal regions are also used to mediate interactions with multiple partners. Terminal regions of flagellin are not only involved in flagellin-flagellin subunit interactions but also in interactions with the hook-associated proteins, HAP3 at the base and HAP2 at the tip, in flagellar assembly in vivo.

Terminal disorder also plays a role in fine-tuning of intersubunit interactions. Conformational entropy of the disordered leg domains of the HAP2 pentamer cap moderates the otherwise too tight HAP2-filament interactions to allow insertion of flagellin subunits below the cap.[41] Disordered terminal regions of flagellin are also involved in sliding and switching interactions essential for polymorphic supercoiling of flagellar filaments.[48] Finally, terminal disorder as a common structural feature of axial proteins seems to serve as the recognition signal for the flagellum-specific export machinery.[49,50]

The Helical Filament

Structural Organization of Flagellin

The flagellar filament is composed of a single protein, flagellin. Flagellin from a wild-type strain of *Salmonella*, SJW1103, is composed of 494 amino acids.[5] The amino acid sequences of the terminal regions of flagellin, including about 180 NH_2-terminal and 100 COOH-terminal residues, are known to be well conserved from species to species of bacteria, while the central region is highly variable.[37,38]

Proteolytic experiments revealed that both terminal regions of flagellin are highly susceptible to even a mild proteolytic treatment. The readily hydrolysed terminal portions span approximately the first 65 and the last 45 residues of the molecule.[39,51] These experiments suggested that there are no significant structural constraints in the terminal regions to limit the accessibility of the peptide bonds, i.e., these regions have no ordered tertiary structure and are in extensive contact with the surrounding media. NMR experiments provided direct evidence for disorder and mobility in the terminal regions of the monomer.[52] In contrast to the poorly ordered terminal regions, the central part of the flagellin molecule is quite stable against proteolysis and has a cooperatively ordered structure.[39] Calorimetric experiments demonstrated that thermal denaturation of the compact part of flagellin is highly reversible and it is composed of three autonomous folding domains.[43]

Far-UV CD and NMR studies demonstrated that the disordered regions become ordered upon filament formation into a largely α-helical domain.[52,53] Electron cryomicroscopy and X-ray diffraction studies have revealed the subunit structure within the filament.[30,54] The overall shape of flagellin looks like an upper case Greek gamma (Γ) with a vertical dimension of about 140 Å and a horizontal dimension of about 110 Å. Flagellin consists of four linearly connected domains labelled D0, D1, D2 and D3, which are arranged from the inside to outside of the filament. The ribbon diagram of the polypeptide backbone in Figure 2 shows the chain folding. The NH2-terminal chain starts from D0, going through D1, D2 and reaches D3 and then comes back through D2 and D1 and the COOH-terminal chain ends in D0. The terminal chains form an α-helical coiled coil in domain D0. Domain D1 comprises an NH_2-terminal segment from Ala 44 to Gln 176 and a COOH-terminal segment from Asn 406 to Glu 454. Domain D2 also comprises two segments: Lys 177 to Gly 189 and Ala 284 to Glu 405. A central segment from Tyr 190 to Val 283 makes up domain D3. All three domain connections are formed by pairs of short antiparallel chains. The one that connects domains D0 and D1 is longer than the other two and therefore it is called the spoke region.

Domain D1 comprises three α-helices and a strand of a unique β-turn/β-hairpin/α-helix motif. This rod-shaped domain forms an extensive hydrophobic core along its central axis in a similar way to four-helix bundles in other protein structures. Because of an irregularity in one of the four strands involving the β-turn/β-hairpin/α-helix in series, however, the packing of hydrophobic side chains is relatively loose in the middle portion. In domain D2, the Cα backbone trace shows

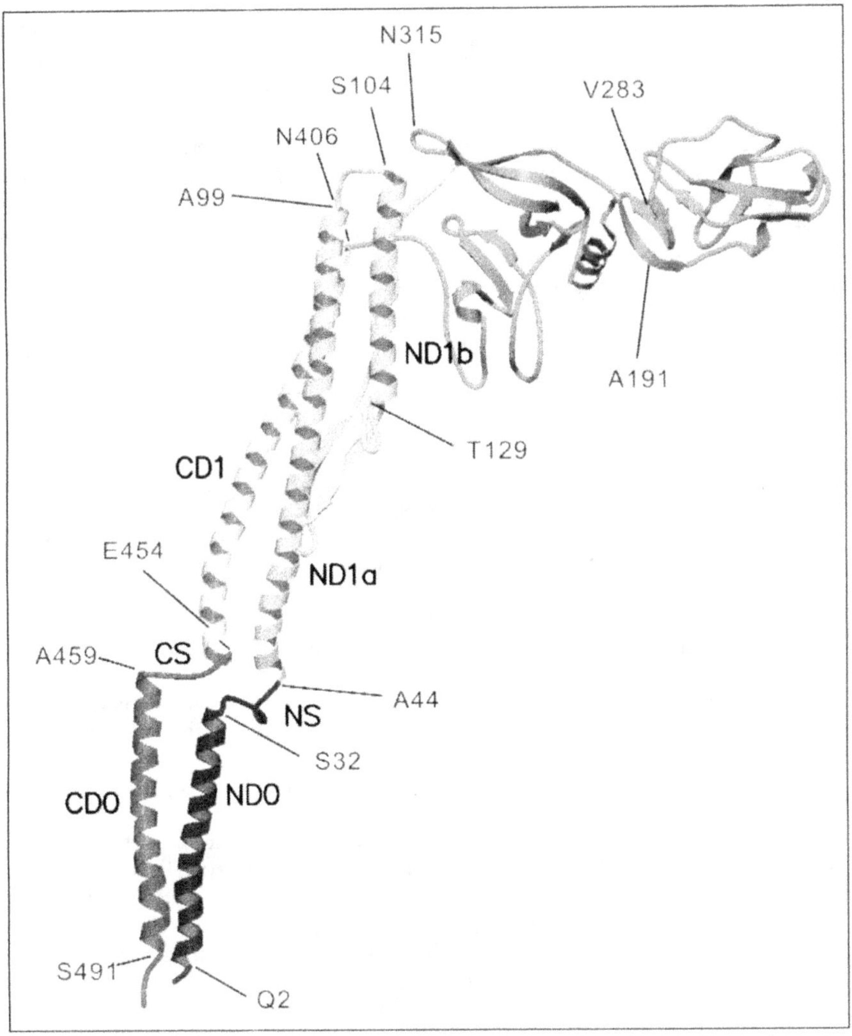

Figure 2. Cα ribbon diagram of flagellin. The rainbow colors from blue to light blue, green, yellow and red represent the sequence from the NH_2- to COOH-terminus. Reprinted with permission from reference 30.

a previously undescribed fold. The fold comprises a series of β-hairpins all pointing away from one another like three thin leaves spread out or like a curve of cubic polynomial called the 'folium of Descartes'. A common feature found in these folds called β-folium is that most of the tips of β-hairpins are either bent or twisted and so some of them are better described as a β-finger with a claw. Flagellin can have domain D3 deleted without losing its ability to form filaments.[55,56] The relatively independent arrangement of this domain from the rest explains why its deletion has a minimal effect on the structural integrity of flagellin.

Filament Structure

The flagellar filament is composed of 11 strands of protofilaments, which are nearly longitudinal arrays of subunits (Fig. 3). The densely packed central core of the filament consists of a concentric double tubular structure made of domains D0 and D1.[30] Domains D2 and D3, which project out from the filament core, are relatively well separated from one another. The diameter of the filament is approximately 230 Å and that of the central channel is about 20 Å. α-helical coiled-coils in domains D0 and D1 stabilize the filament structure by making intimate intersubunit interactions both axially and laterally.[30,57]

The disordered terminal regions are involved in the inner tube of the concentric tubular structure forming long helical bundles and their direct interaction is responsible for the proper folding of the filament core. There is no absolute requirement for either the NH2- or the COOH-terminal disordered region to be present for filament formation; fragments lacking either the whole disordered COOH-terminal region or most of the disordered NH_2-terminal region are still capable of polymerization, although high precipitant concentrations are required for the filament formation and stability.[45] However, both disordered terminal regions cannot be completely removed simultaneously without losing the ability to form the filaments. The basic subunit packing in the filaments made from truncated flagellin fragments was found to be compatible with that of the normal filaments.[46,47] The polymorphic ability of the filaments of the fragments is also largely reduced, but it is not lost completely. Given the result that removal of about 80 of the total 110 disordered terminal residues does not destroy the filament structure, it is reasonable to assume that the main driving force behind filament formation is the interaction of the nondisordered conserved segments. These experiments suggest that the major part of the disordered terminal regions of flagellin is not essential for filament formation, but it does play an important role in stabilizing the filament structures and influence their polymorphic conformation.

Polymorphism of the Helical Filaments

The flagellar filaments may adopt a number of distinct helical forms and can be interconverted between these forms by mechanical force[8] or changes in the environmental conditions.[6,7] Molecular events behind the swimming-tumbling pattern of bacterial motility involve such polymorphic transitions of normally left-handed supercoiled filaments into right-handed forms upon quick reversal of motor rotation that drives the filament.[3,4]

It is not a simple task for chemically identical subunits to construct supercoiled filaments by bonding to each other in a regular manner. Polymorphic supercoiling of the filament is reasonably well explained by a bistable protofilament model in which subunit conformation is assumed to be the same within each protofilament and the protofilament conformation, including interactions with neighboring protofilaments, is classified into two distinct forms.[32,58,59] The model assumes that conformational changes occur cooperatively along the protofilament. When all the protofilaments have an identical conformation the filament morphology is straight. There are actually two types of straight filaments with distinct helical symmetries; one is called the L-type, in which the protofilaments are tilted to the left and the other called the R-type with the protofilaments tilted to the right.[60] The L-type protofilament has a slightly longer intersubunit distance along the protofilament than that of the R-type.[57] If some of the 11 protofilaments are in the L-state and the others in the R-state, this gives rise to curvature, which, together with an inclination of the protofilaments from the filament axis, results in a supercoiled form. When the majority of protofilaments are in the L-type, the resulting supercoils are left-handed. When the number of

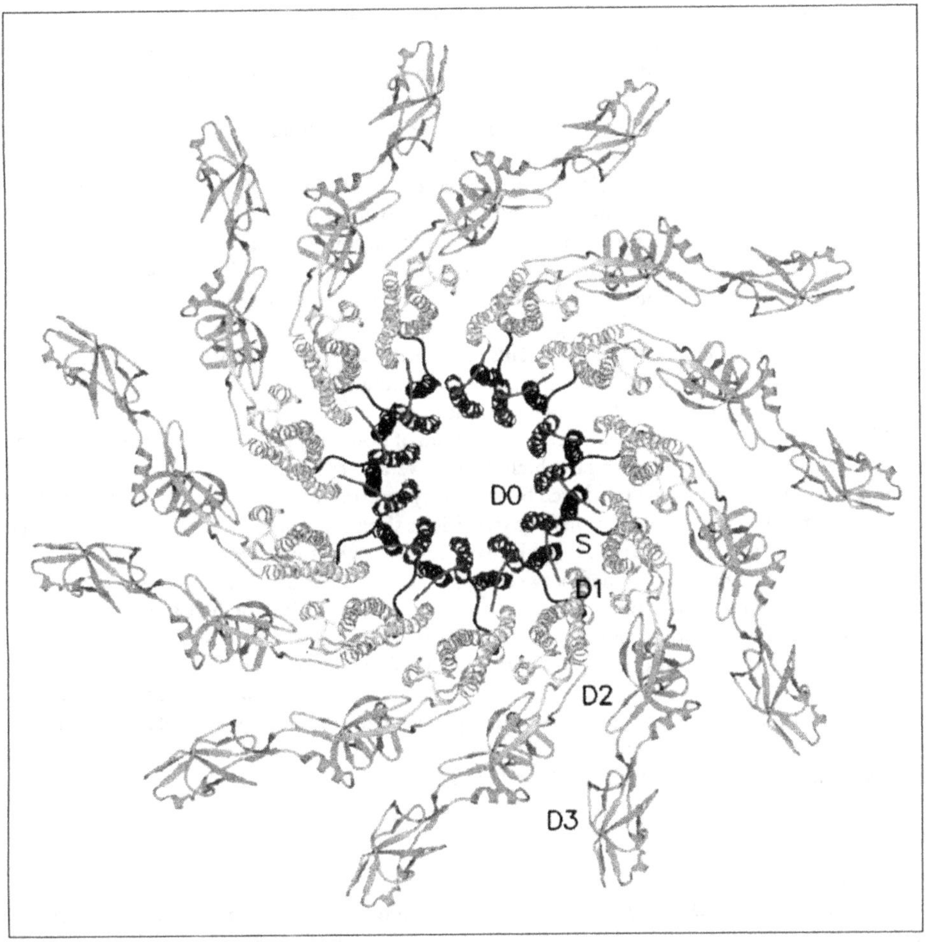

Figure 3. Structure of the flagellar filament viewed from the distal end. Flagellin molecules are shown in the same way as in Figure 2. Reprinted with permission from reference 30.

protofilaments in the R-state increases the resulting supercoils can be right-handed ones due to right-handed inclination of the protofilaments.

The straight filaments isolated from the mutant strains SJW1160 and SJW1655, both of which are point mutants of the wild type strain SJW1103, have the helical symmetry of the L- and R-type respectively[57] and these two types of flagellins can be copolymerized to form different types of supercoiled filaments depending on the mixing ratio.[61] Therefore, the two conformations of these flagellin mutants are thought to represent the two that coexist in the supercoiled filaments.

Although the bistable protofilament model gives a reasonable static and mechanical view of polymorphic supercoiling, it cannot give insights into mechanistic details at the molecular level. Recently, a massive molecular dynamics simulation was performed to further elucidate the detailed underlying molecular mechanisms of the polymorphic supercoiling.[48] The following three types of subunit interactions were found to be essential: "permanent" interactions, which are always maintained between subunits in the various supercoil structures, "sliding" interactions which are formed between variable hydrophilic or hydrophobic residue pairs, allowing intersubunit shear without large changes in energy, while the formation and breakage of "switch" interactions stabilize inter- and intrasubunit interactions, respectively.

Permanent interactions determine the overall architecture of the flagellar filament. Many residues of domain D1 are involved in permanent interactions that are mainly responsible for geometrical constraints of the tubular structure, but many others of domain D1 are involved in sliding and switch interactions. Sliding interactions provide a mechanism that allows large flexibility in the intersubunit interface without large changes in the interaction energy. Switch interactions play essential roles in locking each protofilament interface to be in either the R- or L-type state. The D0 domain was found to be preferentially involved in sliding and switching interactions. Some experimental results suggest that the flagellin monomer tends to be in the R-type conformation if the constraints to form the native tubular structures are weakened. MD simulation showed that the formation of switch salt bridges and hydrogen bonds actually drives the subunit to take the L-type conformation in the filament. This result indicates that switch interactions are formed at the expense of intrasubunit interaction energy, leading to coexistence of the R- and L-type protofilaments in the polymorphic supercoil structures.

MD simulation also shed light on the mechanism of polymorphic transitions from a right- to left-handed supercoil and vice versa, which were driven by the torque applied on domains D0 and D1 to simulate the torque produced by the rotary motor.[48] Here, the transitions clearly show hysteresis. This hysteresis indicates that the transitions should occur in two steps. The initial step mainly involves twist changes while roughly maintaining curvature and accompanying these twist changes, the filament is transformed between right- and left-handed helices. It was observed that the number of switch interactions remained almost constant during the first step. The first step is therefore a geometrical response of the filament helix to external torque. After the first step, the filament structure relaxes slowly into a new metastable state involving changes in switch and other interactions. In this second step, curvature changes while the handedness of the filament is maintained. This two-step mechanism was designated as the "transform and relax" mechanism. More coarse-grained model simulations are under way to understand the dynamic behavior of the polymorphic supercoil transitions in more details.

The Hook

The FlgE Subunit

The polypeptide chain of *Salmonella* hook protein contains 402 amino acid residues. Proteolytic and NMR studies demonstrated that both terminal regions of hook protein, involving about 70 NH_2- terminal and 30 COOH-terminal residues are disordered in solution.[40] Upon polymerization, the disordered terminal regions fold into a predominantly α-helical domain. It has been demonstrated that polymerization ability is lost completely upon removal of both disordered terminal regions.[44]

Attempts to crystallize intact hook protein have been unsuccessful since it readily assembles into filamentous structures under crystallization conditions, instead of forming 3D crystals. However, the FlgE31 fragment corresponding to residues 71-369 out of 402 and having a molecular mass of 31 kDa was successfully expressed in *E. coli*, purified and crystallized. This fragment lacks both terminal regions that are unfolded in the monomeric form in solution.[44] The Cα backbone trace of FlgE31 (Fig. 4) obtained by X-ray crystallography shows two domains, D1 and D2, connected by a short stretch of two-stranded anti-parallel β-sheet.[62] The longest dimensions of domain D1 and D2 are about 50 Å and 45Å, respectively and these two domains are connected along their long axes with an angle of about 70°. The amino-terminal segment from Gly 71 to Ala 144 and the carboxy-terminal segment from Pro 285 to Ser 363 comprise D1 and a central segment from Ala 145 to Lys 284 makes up domain D2. Both domains have an oval shape and are made mostly of β-structures. Domain D2 is a flattened, eight-strand β-barrel but with significant irregularity and extra loops. Domain D1 has a rather complex, unique fold composed of many different folding motifs: a stack of four β-hairpins one above another, alternating their orientations with

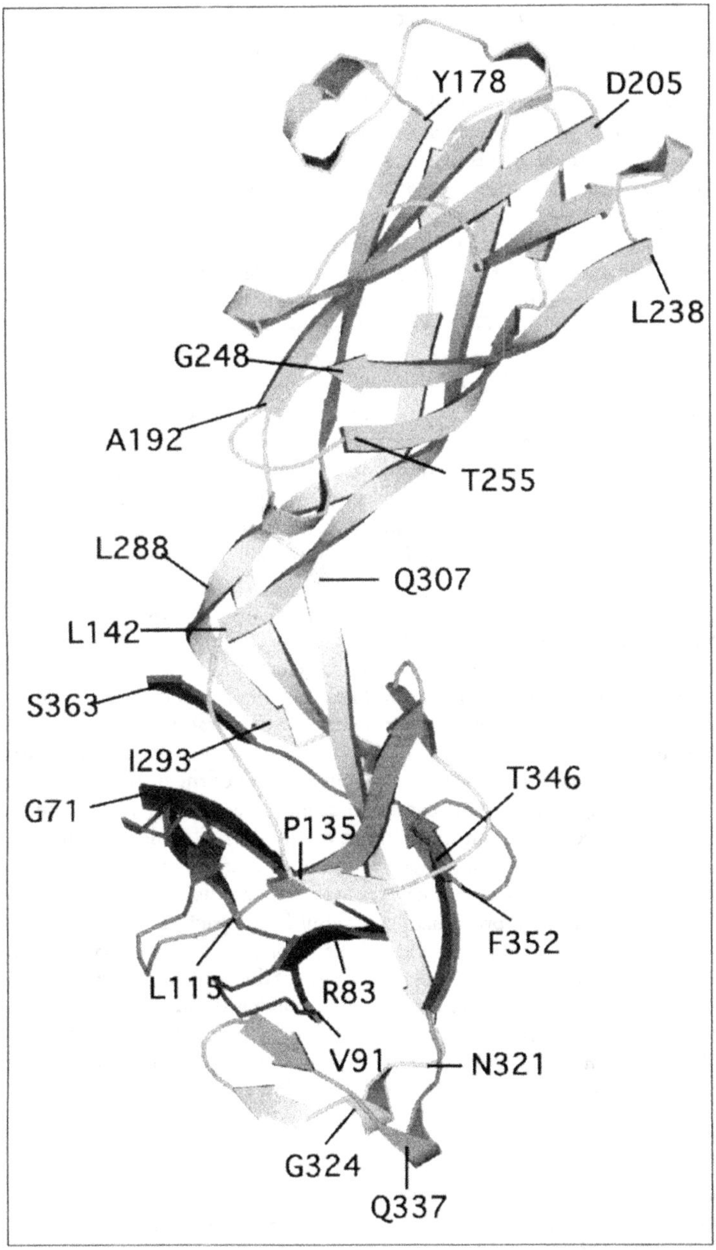

Figure 4. Cα ribbon diagram of FlgE31, a 31 kDa fragment of the hook protein FlgE missing both terminal regions. The sequence from the NH_2- to COOH-terminus is represented in the same way as in Figure 2. Reprinted with permission from reference 62.

crossing angles of about 120°; a triangular loop; a four-stranded and a two-stranded β-sheet; two consecutive β-turns; and a vertically extended chain, which appears to be the backbone around which the other motifs assemble.

Structure and Polymorphic Supercoiling of Hook

The bacterial flagellar hook acts as a molecular universal joint, transmitting torque produced by the rotary motor to the filament. The hook is much more flexible than the filament.[63] The hook, like the helical filament, is also an assembly of 11 circularly arranged protofilaments and shows polymorphic supercoiling, which is only visible in polyhook structures.[11] A specific aspect of the hook structure is its very small radius of curvature and short helical pitch. The most plausible mechanical explanation of how the hook works as a universal joint is that a supercoiled form is converted repeatedly into an identical form by permutation of the protofilament conformations from one to the next along the circumference in the direction opposite to its rotation, keeping the helical axis of the supercoil fixed in the same position and orientation.

It is curious how hook protein and flagellin with very different structural characteristics form tubular structures with basically the same architecture and helical symmetry. Low resolution EM analysis revealed that the hook subunit is composed of three morphological domains: the rod shaped inner domain (D1), the middle spherical domain (D2) and the outer ellipsoidal domain.[64] The outer domain, D2, hangs out off the middle domain D1. There is an axial gap between the D2 domains, which provides freedom for the hook to curve. On the inner side of D1 are the rod-like features that form the inner tube with a narrow central channel. This design allows the structure to bend without breaking. In the case of the hook, the design represents a way to maintain structural integrity while still permitting a more flexible and more highly curved structure than the filament, which has a more rigid, double-tube design. The central channel of the hook has a diameter similar to that of the filament.

Higher resolution structural information for the hook was obtained by fitting the atomic structure of the FlgE31 fragment determined by X-ray crystallography to a density map of a straight hook reconstructed from images obtained by electron cryomicroscopy (Fig. 5).[65] The two domains of FlgE31 were docked into the middle and outer domains of the map. The missing terminal sequences, which are unfolded in the monomer and are stabilized into α-helical bundles upon polymerization, were supposed to form the innermost rod-shaped domain in the hook.

Based on the obtained structure, the molecular mechanism of the universal joint function of the hook has been investigated by a ~ two-million-atom molecular dynamics simulation.[66] On the inner side of the supercoil, protein subunits are highly packed along the protofilament and no gaps remain for further compression, whereas subunits are largely apart and slightly separated with hydrogen bonding through one layer of water molecules on the outer side (Fig. 6). As for the intersubunit interactions between protofilaments, subunits are packed along the 6-start helix in a left-handed supercoil whereas they are highly packed along the 5-start helix in a right-handed supercoil. It was concluded that the supercoiled structures of the hook in the left- and right-handed forms make maximal use of the gaps between subunits, which was called as the gap compression/extension mechanism. Mutual sliding of subunits at the subunit interface accompanying rearrangements of intersubunit hydrogen bonds is interpreted as a mechanism to allow continuous structural change of the hook during flagellar rotation at low energy cost.

The HAP2 Cap

HAP2 Monomer and Self-Association

HAP2 appears to be a distinct member among the axial proteins, which does not show any significant sequential similarity to other components of the flagellar system.[36] Inspite of its dissimilarity, HAP2 also possesses disordered terminal regions like flagellin and hook protein. HAP2 from *Salmonella* SJW1103 is composed of 466 amino acid residues and its highly mobile parts span about 42 NH_2-terminal and 51 COOH-terminal residues.[41] Secondary structure prediction for *Salmonella* HAP2 suggested that the terminal regions of HAP2 are dominated by α-helical structure, which is a general characteristic for the flagellar axial proteins. Removal of the disordered terminal portions by trypsin results in a fragment of 40 kDa (HP40), which is quite stable against proteolysis, indicating that it is composed of compact structural units. Unlike flagellin and hook

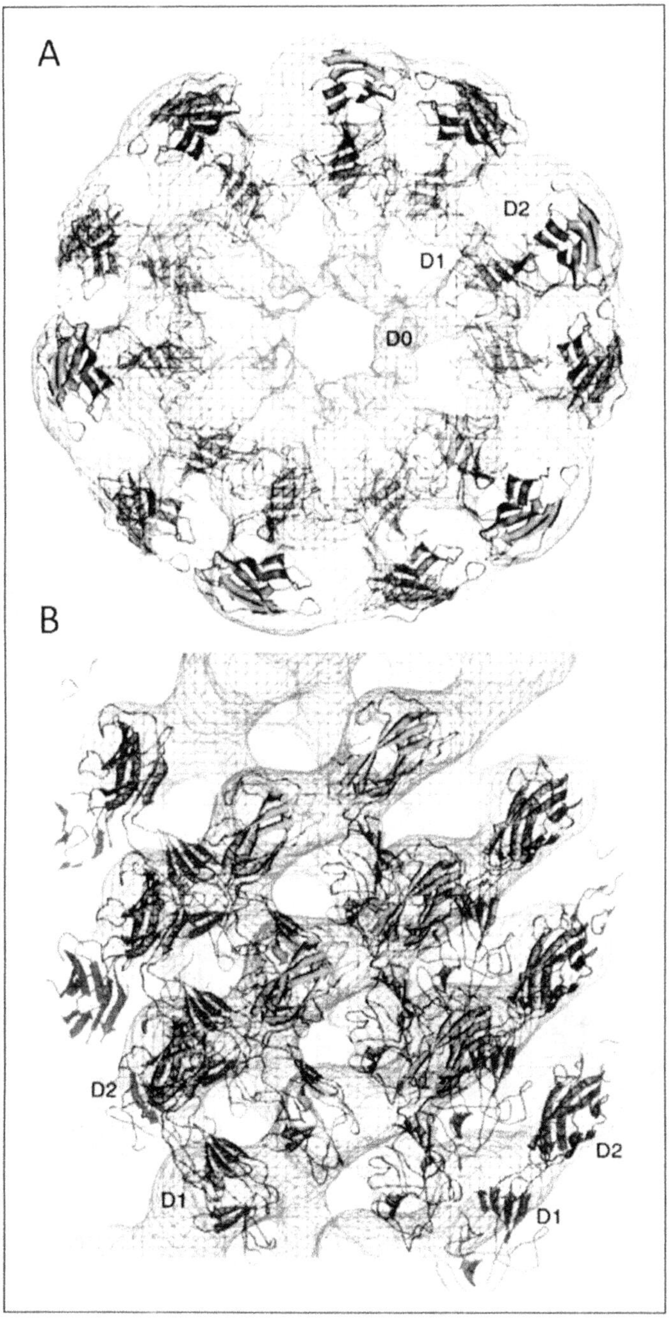

Figure 5. Molecular models of FlgE31 in Cα ribbon diagram docked in the density map of a straight hook obtained by electron cryomicroscopy and image analysis. Upper panel, view from the distal end; lower panel, side view. FlgE31 consists of domains D1 and D2. The density for domain D0 is not filled in because FlgE31 is missing both terminal regions. Reprinted with permission from reference 62.

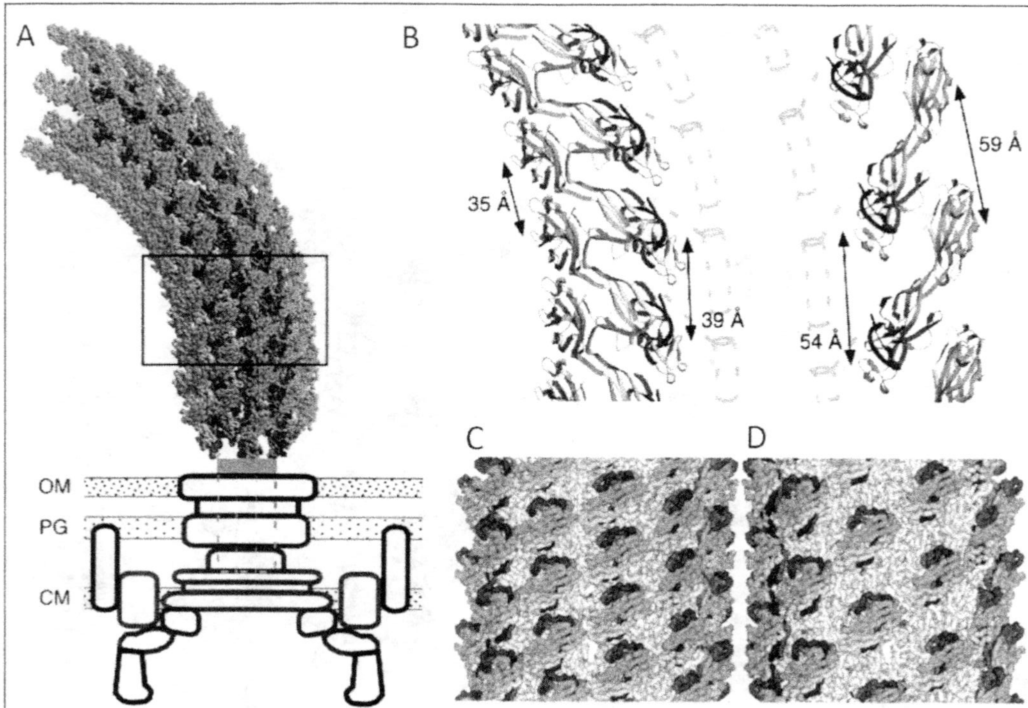

Figure 6. Atomic model of supercoiled hook. A) Atomic model of coiled hook with a schematic diagram of the basal body spanning the outer (OM) and cytoplasmic (CM) membranes as well as the peptidoglycan layer (PG). This coiled hook model is part of supercoiled polyhook with a helical pitch of 950 Å and a diameter of 350 Å. B) Magnified image of the coiled hook with the innermost and outermost protofilaments on the left and right, respectively. The inner core domains made of both terminal chains and the central channel are represented by dotted grey lines. C) Intermolecular packing arrangements of D2 domains on the inner side and (D) on the outer side of the coiled hook surface. Only domain D2 is colour coded as in Figure 1, while domain D1 is colored light gray. Reprinted with permission from reference 62.

protein, which have a central part composed of β-structural domains, the central portion of the HAP2 molecule appears to contain comparable amounts of α and β-structure.[41]

The purified HAP2 protein has been found to readily self-assemble into annular structures in vitro, which are composed of ten subunits and have a pentagonal shape.[67,68] The HP40 fragments do not form the decamer, but they can assemble into the pentamer.[41] The disordered terminal regions of monomeric HAP2 are stabilized and protected in the decameric form. The three-dimensional structure reconstructed from electron micrographs of negatively stained HAP2 decamers showed a bipolar pair of pentamers (Fig. 7A), where each pentamer was composed of a pentagonal plate and five leg domains.[69]

HAP2 performs its physiological capping function as a pentamer.[69,70] The disordered terminal regions are not essential for the formation of HAP2 pentamers, which is governed by interactions between the compact central portions of the subunits.[41] The terminal regions of HAP2 in the pentamers are exposed and disordered, like in the case of monomeric HAP2 and are readily degraded by proteases. It seems that the major role of the highly conserved and mobile terminal regions is to participate in capping interactions and to moderate the affinity of the HAP2 cap for the filament by imposing an entropic penalty for binding.

Structure of the Cap-Flagellar Filament Complex

The three-dimensional density map of the cap-filament complex was reconstructed by single particle image analysis (Fig. 7B).[69] There is a lid-like plate or cap at the top of filaments which is 120 Å wide and 25 Å thick; the diameter roughly corresponds to that of the outer tube of the filament. Its pentagonal shape in the end-on view indicates that the cap is a pentamer of HAP2. Inside the filament just below the cap plate, there is a cavity roughly 40 Å wide and 70 Å deep. Presumably, flagellin subunits are exported through the narrow central channel in a partially unfolded state. The cavity appears to have the right size to accommodate only one flagellin subunit at a time, allowing its refolding without aggregation with other subunits.

The cap is attached to the distal end of the filament at five positions through its anchor domains, where these anchor domains show significant deviation from the fivefold symmetry. The five anchor domains are well separated from one another, forming five gaps with different sizes and shapes between the plate and the filament end. One of the gaps is distinctly larger than the other four, having an inverted L shape and its size corresponds to the size of domain D1 of flagellin. This is most likely to be the site of flagellin assembly.

Mechanism of Flagellin Assembly

Because of the axial stagger by half a subunit in the lateral subunit packing, the end of the filament is not flat but has five indentations around its circumference, where one of them is a double indentation. It appears that four of the five indentations are filled with the HAP2 leg domains, while the one with double indentation is not filled by the last leg domain.[69,70] This leaves an inverted L-shaped opening fully open for flagellin assembly. This opening has the right size and shape and probably an appropriate atomic surface structure, for domain D1 of the newly arriving flagellin subunit to insert and bind.

There is a mismatch in the rotational symmetry between the cap and the filament structures. While the 11 protofilaments are arranged into the tubular structure of the filament with the axial stagger to have a nearly 5.5-fold circular symmetry, the cap has a 5-fold circular symmetry. Because of this symmetry mismatch between the filament and the cap, the five leg domains of the cap must have significantly different conformations from one another for interaction with flagellin subunits.[41,71] The highly flexible nature of the leg domain, as partly indicated by the disordered terminal regions in the isolated cap, allows such polymorphic interactions, reducing conformational strains and permitting stable interactions between the two structures in spite of the symmetry mismatch. Upon every subunit incorporation, four legs of the cap rearrange their conformations and the last one changes its binding partner; namely these legs walk along the helical steps of the filament end.[69,71] Every insertion of a flagellin subunit is likely to force the cap to move into the next stable position energetically equivalent to the current position. The cap moves up 4.7 Å and

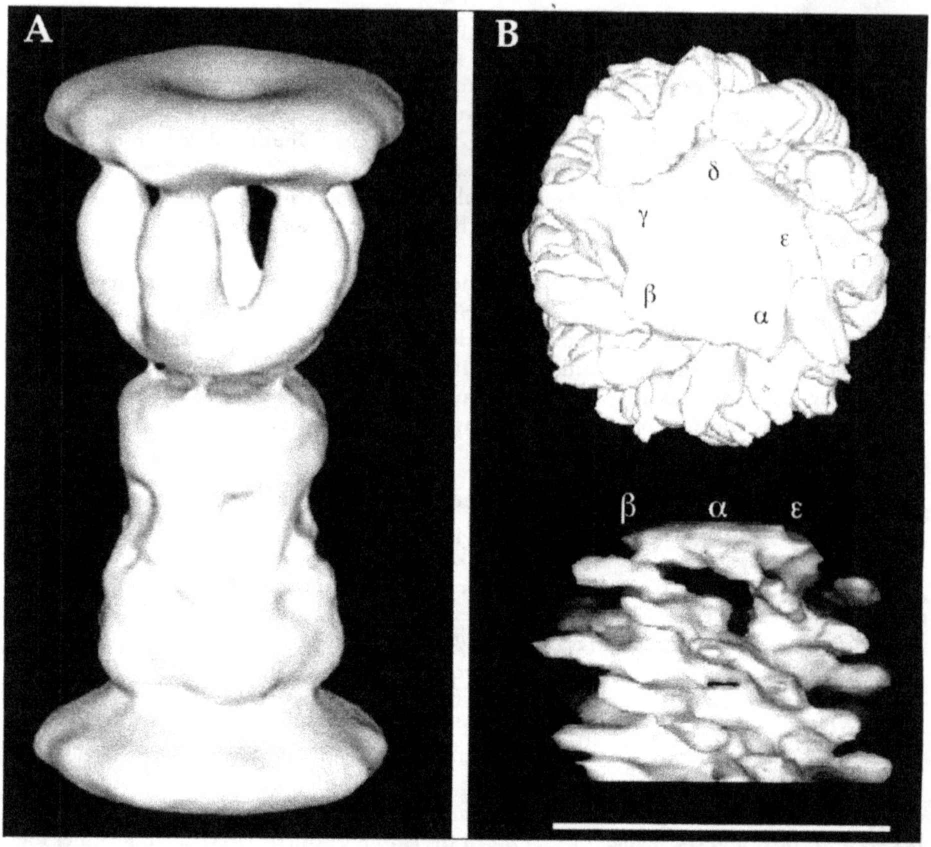

Figure 7. A) Solid surface representation of the three-dimensional density map of the HAP2 decamer. The diameter is 120 Å. Two pentamer caps (top and bottom), each composed of a plate and five leg domains, are connected with a connector domain at the center, which is composed of both terminal regions that are disordered in the pentameric or monomeric form of HAP2 in solution. Only the upper pentamer was aligned and used for image reconstruction, because the connector domain has a significant azimuthal flexibility and therefore two pentamers are not well registered to each other. That is why the lower pentamer shows only a blurred image. B) Three-dimensional density map of the cap-filament complex. Upper panel shows an end-on view from the distal end, showing a pentagonal shape of the plate domain of the cap. Lower pannel is a side view showing a regular helical array of flagellin subunits and the plate domain of the cap. The inverted L-shaped gap between the cap plate and the filament is likely to be the assembly site for the newly arriving flagellin subunit. Bar represents the diameter of the filament, 230 Å. (Adapted with permission from ref. 69).

rotates 6.5 degree along the left-handed 5-start helix of the filament. This would result in a roughly complete rotation of the cap by the assembly of 55 flagellin subunits. The energy for the cap rotation with the conformational rearrangements is presumably supplied by the binding energy of the newly incorporated flagellin subunit at the distal end of the filament.[41] The binding of flagellin subunits are due to their high local concentration. The flagellar type III protein export apparatus presumably located on the cytoplasmic face of the basal body translocates flagellin subunits into the export channel to keep the effective concentration of flagellin at the distal end of the export channel above the critical concentration. The energy source for the flagellar protein export is maily proton motive force with a help of FliI ATPase to make the export process more efficient.[83] The energy transduction network formed here is unique and interesting, because the energy produced in the cytoplasmic space is transmitted quite a long distance (10-15 μm) through flagellin export to be used for the helical movement of the cap at the distal growing end.

Applications

Flagella Display Technology

Kuwajima et al[56] reported that various deletions can be introduced into the hypervariable central region of *E. coli* flagellin (497 residues) without destroying its ability to form the flagellar filaments. The smallest deletion mutant lacked 187 residues and retained only the NH_2-terminal 193 residues and the COOH-terminal 117 residues. Since then, the surface-exposed dispensable central region of the *fliC* gene has been the target position for insertions of foreign peptide/protein gene sequences to be displayed on flagella. The use of recombinant microorganisms with surface exposed protein structures has gained a lot of interest during the last decade since this type of research holds great promise in different areas such as biotechnology, microbiology and biomedicine.

The most frequent application of flagella display is the construction of novel recombinant vaccines.[72-75] Live bacterial vaccine delivery vehicles have been developed through the surface display of foreign antigens on the flagellar filament surface. Flagella display has also been used for the display of a random peptide library as an alternative to the phage display technique.[76,77] Lu et al[78] constructed a fusion protein (FliTrx) in which the entire coding sequence of *E. coli* thioredoxin was inserted into the dispensable region of the flagellin gene *fliC*. Flagella display has proven to be applicable also in bacterial adhesion technology since large fragments of bacterial adhesins up to 302 amino acid residues can be functionally expressed as fusions to flagellin.[79]

Flagella Nanotubes

Kumara et al[80-81] demonstrated the versatility of employing bioengineered flagella for the generation of a variety of nanoparticle arrays and nanotubes. The FliTrx fusion protein was genetically engineered to display rationally designed histidine, arginine-lysine and aspartic acid-glutamic acid peptide loops on the solvent-accessible outer domain region. The resulting flagellin monomers were polymerized to obtain corresponding oligomeric flagellar bionanotubes in which the peptide loops were 5 nm apart along each protofilament. These flagella nanotubes were able to bind various metal ions. Controlled reduction of these metal ions yielded ordered arrays of nanoparticles and nanotubes. Covalent attachment of separately synthesized gold nanoparticles to the flagella scaffold was also demonstrated. Flagellar nanotubes were also used to initiate the formation of silica and titania bionanotubes.

Also, *E. coli* flagellin was functionalized to form tailored nanotubes that bind single types or pairs of ligands, including divalent cations, fluorescent antibodies, or biotin-avidin-linked moieties such as ferritins.[82] Tobacco Etch Virus (TEV) protease site-containing flagellins were cleaved by the cognate protease without filament disintegration, potentiating their use as removable nanolithography masks to deposit attached ligands by protease cleavage.

Conclusion

Formation of complex biological structures such as cells and cellular organelles is based on self-assembly of component macromolecules such as proteins and nucleic acids either in solution or within lipid membranes. The self-assembly processes are all driven by precise recognition of template structures by assembling molecules, both of which have well-defined three-dimensional atomic arrangements for specific binding and interactions, and yet extensive flexibility of these molecules including disordered conformation plays essential roles in formation and stabilization of complex structures that are otherwise impossible to form. The bacterial flagellum is one of the typical examples as described in this article. The well-established abilities of biological macromolecules to self-organize into complex three-dimensional architectures therefore give us a wonderful opportunity to use them in nanotechnology applications such as design of self-assembling nanostructures for nano-devices and nano-systems.

Acknowledgements

We thank F. Oosawa, S. Asakura, H. Hotani, P. Závodszky, F. Friedler and D.L.D. Caspar for continuous support and encouragement. This work was supported in part by Grants-in-Aid for Scientific Research to K.N. (16087207) from the Ministry of Education, Science and Culture of Japan.

References

1. Macnab RM. How bacteria assemble flagella. Annu Rev Microbiol 2003; 57:77-100.
2. Namba K, Vonderviszt F. Molecular architecture of bacterial flagellum. Q Rev Biophys 1997; 30:1-65.
3. Macnab RM, Ornston MK. Normal-to-curly flagellar transitions and their role in bacterial tumbling. Stabilization of an alternative quaternary structure by mechanical force. J Mol Biol 1977; 112:1-30.
4. Turner L, Ryu WS, Berg HC. Real-time imaging of fluorescent flagellar filaments. J Bacteriol 2000; 182:2793-2801.
5. Kanto S, Okino H, Aizawa SI et al. Amino acids responsible for flagellar shape are distributed in terminal regions of flagellin. J Mol Biol 1991; 219:471-480.
6. Kamiya R, Asakura S. Helical transformations of Salmonella flagella in vitro. J Mol Biol 1976; 106:167-186.
7. Kamiya R, Asakura S. Flagellar transformations at alkaline pH. J Mol Biol 1977; 108:513-518.
8. Hotani H. Micro-video study of moving bacterial flagellar filaments. III. Cyclic transformation induced by mechanical force. J Mol Biol 1982; 156:791-806.
9. DeRosier DJ. Spinning tails. Curr Opin Struct Biol 1995; 187-193.
10. Hirano T, Yamaguchi S, Oosawa K et al. Roles of FliK and FlhB in determination of flagellar hook length in Salmonella typhimurium. J Bacteriol 1994; 176:5439-5449.
11. Kagawa H, Aizawa SI, Asakura S. Transformations in isolated polyhooks. J Mol Biol 1979; 129:333-336.
12. Saijo-Hamano Y, Uchida N, Namba K et al. In vitro characterization of FlgB, FlgC, FlgF, FlgG and FliE, flagellar basal body proteins of Salmonella. J Mol Biol 2004; 339:423-435.
13. Ikeda T, Homma M, Iino T et al. Localization and stoichiometry of hook-associated proteins within Salmonella typhimurium flagella. J Bacteriol 1987; 169:1168-1173.
14. Ikeda T, Asakura S, Kamiya R. Cap on the tip of Salmonella flagella. J Mol Biol 1985; 184:735-737.
15. Macnab RM. Type III flagellar protein export and flagellar assembly. BBA 2004; 1694:207-217.
16. Blocker A, Komoriya K, Aizawa S-I. Type III secretion systems and bacterial flagella: insights into their function from structural similarities. Proc Natl Acad Sci USA 2003; 100:3021-3030.
17. Berg HC. The rotary motor of bacterial flagella. Annu Rev Biochem 2003; 72:19-54.
18. Kojima S, Blair DF. The bacterial flagellar motor: structure and function of a complex molecular machine. Int Rev Cytol 2004; 233:93-134.
19. Bourret RB, Stock AM. Molecular information processing: lessons from bacterial chemotaxis. J Biol Chem 2002; 277:9625-9628.
20. Yonekura K, Maki-Yonekura S, Namba K. Growth mechanism of the bacterial flagellar filament. Res Microbiol 2002; 153:191-197.
21. Kubori T, Shimamoto N, Yamaguchi S et al. Morphological pathway of flagellar assembly in Salmonella typhimurium. J Mol Biol 1992; 226:433-446.
22. Fan F, Macnab RM. Enzymatic characterization of FliI: an ATPase involved in flagellar assembly in Salmonella typhimurium. J Biol Chem 1996; 271:31981-31988.
23. Nambu T, Minamino T, Macnab RM et al. Peptidoglycan-hydrolyzing activity of the FlgJ protein, essential for flagellar rod formation in Salmonella typhimurium. J Bacteriol 1999; 181:1555-1561.
24. Onishi K, Ohta Y, Aizawa S-I et al. FlgD is a scaffolding protein needed for flagellar hook assembly in Salmonella typhimurium. J Bacteriol 1994; 176:2272-2281.
25. Williams AW, Yamaguchi S, Togashi F et al. Mutations in FliK and FlhB affecting flagellar hook and filament assembly in Salmonella typhimurium. J Bacteriol 1996; 178:2960-2970.
26. Shibata S, Takahashi N, Chevance FF et al. FliK regulates flagellar hook length as an internal ruler. Mol Microbiol 2007; 64:1404-1415.
27. Moriya N, Minamino T, Hughes KT et al. The type III flagellar export specificity switch is dependent on FliK ruler and a molecular clock. J Mol Biol 2006; 359:466-477.
28. Ferris HU, Minamino T. Flipping the switch: bringing order to flagellar assembly. Trends Microbiol 2006; 14:519-526.
29. Iino T. Assembly of Salmonella flagellin in vitro and in vivo. J Supramol Str 1974; 2:372-384.
30. Yonekura K, Maki-Yonekura S, Namba K. Complete atomic model of the bacterial flagellar filament by electron cryomicroscopy. Nature 2003; 424:643-650.

31. Ikeda T, Yamaguchi S, Hotani H. Flagellar growth in a filament-less Salmonella fliD mutant supplemented with purified hook-associated protein 2. J Biochem 1993; 114:39-44.
32. Asakura S. Polymerization of flagellin and polymorphism of flagella. Adv Biophys 1970; 1:99-155.
33. Kato S, Aizawa SI, Asakura S. Reconstruction in vitro of the flagellar polyhook from Salmonella. J Mol Biol 1982; 161:551-560.
34. Ikeda T, Asakura S, Kamiya R. Total reconstitution of Salmonella flagellar filaments from hook and purified flagellin and hook-associated proteins. J Mol Biol 1989; 209:109-114.
35. Homma M, Kutsukake K, Hasebe M et al. FlgB, FlgC, FlgF and FlgG. A family of structurally related proteins in the flagellar basal body of Salmonella typhimurium. J Mol Biol 1990; 211:465-477.
36. Homma M, DeRosier DJ, Macnab RM. Flagellar hook and hook-associated proteins of Salmonella typhimurium and their relationship to other components of the flagellum. J Mol Biol 1990; 213:819-832.
37. Wilson DR, Beveridge TJ. Bacterial flagellar filaments and their component flagellins. Can J Microbiol 1994; 39:67-71.
38. Beatson SA, Minamino T, Pallen MJ. Variation in bacterial flagellins: from sequence to structure. Trends Microbiol 2006; 14:151-155.
39. Vonderviszt F, Kanto S, Aizawa SI et al. Terminal regions of flagellin are disordered in solution. J Mol Biol 1989; 209:127-133.
40. Vonderviszt F, Ishima R, Akasaka K et al. Terminal disorder: a common structural feature of the axial proteins of bacterial flagellum? J Mol Biol 1992; 226:575-579.
41. Vonderviszt F, Imada K, Furukawa Y et al. Mechanism of self-association and filament capping by flagellar HAP2. J Mol Biol 1998; 284:1399-1416.
42. Furukawa Y, Imada K, Vonderviszt F et al. Interactions between bacterial flagellar axial proteins in their monomeric state in solution. J Mol Biol 2002; 318:889-900.
43. Vonderviszt F, Uedaira H, Kidokoro SI. Structural organization of flagellin. J Mol Biol 1990; 214:97-104.
44. Vonderviszt F, Zavodszky P, Ishimura M et al. Structural organization and assembly of flagellar hook protein from Salmonella typhimurium. J Mol Biol 1995; 251:520-532.
45. Vonderviszt F, Aizawa SI, Namba K. Role of the disordered terminal regions of flagellin in filament formation and stability. J Mol Biol 1991; 221:1461-1474.
46. Mimori-Kiyosue Y, Vonderviszt F, Namba K. Locations of terminal segments of flagellin in the filament structure and their roles in polymerization and polymorphism. J Mol Biol 1997; 270:222-237.
47. Mimori-Kiyosue Y, Vonderviszt F, Yamashita I et al. Direct interaction of flagellin termini essential for polymorphic ability of flagellar filament. Proc Natl Acad Sci USA 1996; 93:15108-15113.
48. Kitao A, Yonekura K, Maki-Yonekura S et al. Switch interactions control energy frustration and multiple flagellar filament structures. Proc Natl Acad Sci USA 2006; 103:4894-4899.
49. Vegh BM, Gal P, Dobo J et al. Localization of the flagellum-specific secretion signal in Salmonella flagellin. Biochem Biophys Res Commun 2006; 345:93-98.
50. Kornacker MG, Newton A. Information essential for cell-cycle-dependent secretion of the 591-residue Caulobacter hook protein is confined to a 21-amino-acid sequence near the N-terminus. Mol Microbiol 1994; 14:73-85.
51. Kostyukova AS, Pyatibratov MG, Filimonov VV et al. Flagellin parts acquiring a regular structure during polymerization are disposed on the molecule ends. FEBS Letters 1988; 241:141-144.
52. Aizawa SI, Vonderviszt F, Ishima R et al. Termini of Salmonella flagellin are disordered and become organized upon polymerization into flagellar filament. J Mol Biol 1990; 211:673-677.
53. Uratani Y, Asakura S, Imahori K. A circular dichroism study of Salmonella flagellin: evidence for conformational change on polymerization. J Mol Biol 1972; 85-98.
54. Samatey FA, Imada K, Nagashima S et al. Structure of the bacterial flagellar protofilament and implications for a switch for supercoiling. Nature 2001; 410:331-337.
55. Mimori-Kiyosue Y, Yamashita I, Fujiyoshi Y et al. Role of the outermost subdomain of Salmonella flagellin in the filament structure revealed by electron cryomicroscopy. J Mol Biol 1997; 270:222-237.
56. Kuwajima G. Construction of a minimum-size functional flagellin of Eschericia coli. J Bacteriol 1988; 170:3305-3309.
57. Yamashita I, Hasegawa K, Suzuki H et al. Structure and switching of bacterial flagellar filaments studied by X-ray fiber diffraction. Nat Struct Biol 1998; 5:125-132.
58. Calladine CR. Construction of bacterial flagella. Nature 1975; 255:121-124.
59. Calladine CR. Change of waveform in bacterial flagella—role of mechanics at molecular level. J Mol Biol 1978; 118:457-479.
60. Kamiya R, Asakura S, Wakabayashi K et al. Transition of bacterial flagella from helical to straight forms with different subunit arrangements. J Mol Biol 1979; 131:725-742.

61. Kamiya R, Asakura S, Yamaguchi S. Formation of helical filaments by copolymerization of two types of 'straight' flagellins. Nature 1980; 286:628:630.
62. Samatey FA, Matsunami H, Imada K et al. Structure of the bacterial flagellar hook and implication for the molecular universal joint mechanism. Nature 2004; 431:1062-1068.
63. Kato S, Okamoto M, Asakura S. Polymorphic transition of the flagellar polyhook from Eschericia coli and Salmonella typhimurium. J Mol Biol 1984; 173:463-476.
64. Morgan DG, Macnab RM, Francis NR et al. Domain organization of the subunit of the Salmonella typhimurium flagellar hook. J Mol Biol 1993; 229:79-84.
65. Shaikh TR, Thomas DR, Chen JZ et al. A partial atomic structure for the flagellar hook of Salmonella typhimurium. Proc Natl Acad Sci USA 2005; 102:1023-1028.
66. Furuta T, Samatey FA, Matsunami H et al. Gap compression/extension mechanism of bacterial flagellar hook as the molecular universial joint. J Struct Biol 2007; 157:481-490.
67. Ikeda T, Oosawa K, Hotani H. Self-assembly of the filament capping protein, FliD, of bacterial flagella into an annular structure. J Mol Biol 1996; 259:679-686.
68. Imada K, Vonderviszt F, Furukawa Y et al. Assembly characteristics of flagellar cap protein HAP2 of Salmonella: decamer and pantamer in the pH-sensitive equilibrium. J Mol Biol 1998; 277:883-891.
69. Yonekura K, Maki S, Morgan DG et al. The bacterial flagellar cap as the rotary promoter of flagellin self-assembly. Science 2000; 290:2148-2152.
70. Maki S, Imada K, Furukawa Y et al. Plugging interactions of HAP2 pentamer into the distal end of flagellar filament revealed by electron microscopy. J Mol Biol 1998; 277:771-777.
71. Maki-Yonekura S, Yonekura K, Namba K. Domain movements of HAP2 in the cap-filament complex formation and growth process of the bacterial flagellum. Proc Natl Acad Sci USA 2003; 100:15528-15533.
72. Wu JY, Newton S, Judd A et al. Expression of immunogenic epitopes of hepatitis B surface antigen with hybrid flagellin proteins by a vaccine strain of Salmonella. Proc Natl Acad Sci USA 1989; 86:4726-4730.
73. Newton SMC, Joys TM, Anderson SA. Expression and immunogenicity of an 18-residue epitope of HIV1 gp41 inserted in the flagellar protein of a Salmonella live vaccine. Res Microbiol 1995; 146:203-216.
74. Stocker BA, Newton SM. Immune responses to epitopes inserted in Salmonella flagellin. Int Rev Immunol 1994; 11:167-178.
75. Cattozo EM, Stocker BAD, Radaelli A et al. Expression and immunogenicity of V3 loop epitopes of HIV-1, isolates SC and WMJ2, inserted in Salmonella flagellin. J Biotech 1997; 56:191-203.
76. Westerlund-Wikstrom B. Peptide display on bacterial flagella:principles and applications. Int J Med Microbiol 2000; 290:223-230.
77. Westerlund-Wikstrom B, Tanskanen J, Virkola R et al. Functional expression of adhesive peptides as fusions to Eschericia coli flagellin. Prot Eng 1997; 10:1319-1326.
78. Lu Z, Murray KS, Van Cleave V et al. Expression of thioredoxin random peptide libraries on the Eschericia coli cell surface as functional fusions to flagellin: a system designed for exploring protein-protein interactions. Biotechnology 1995; 13:366-372.
79. Tanskanen J, Korhonen TK, Westerlund-Wikstrom B. Construction of a multihybrid display system: flagellar filaments carrying two foreign adhesive peptides. Appl Env Microbiol 2000; 66:4152-4156.
80. Kumara MT, Muralidharan S, Tripp BC. Generation and characterization of inorganic and organic nanotubes on bioengineered flagella of mesophilic bacteria. J Nanosci Nanotechnol 2007; 7:2260-2272.
81. Kumara MT, Srividya N, Muralidharan S et al. Bioengineered flagella protein nanotubes with syteine loops: self-assembly and manipulation in an optical trap. Nano Lett 2006; 6:2121-2129.
82. Woods RD, Takahashi N, Aslam A et al. Bifunctional nanotube scaffolds for diverse ligands are purified simply from Eschericia coli strains coexpressing two functionalized flagellar genes. Nano Lett 2007; 7:1809-1816.
83. Minamino T, Namba K. Distinct roles of the FliI ATPase and proton motive force in bacterial flagellar protein export. Nature 2008; 451:485-88.

Chapter 6

Keratin Intermediate Filaments:
Similarities and Differences with Other Members of the IF Family

David A.D. Parry* and Thomasin A. Smith

Abstract

Intermediate filament (IF) molecules are characterised by a near-common, two-stranded α-helical rod domain containing four coiled-coil segments and three linker regions. The structures of parts of the coiled-coil segments (but not that of the linkers) have been determined at high resolution. In contrast to the commonality of the rod domain there are regions of sequence that are N- and C-terminal to the rod (designated head and tail respectively) that differ markedly in chemical character from one type of IF to another. Little is currently known about their tertiary conformations and, indeed, it is only for the nuclear lamin IF molecules that a terminal domain has been crystallized and its structure solved at atomic resolution. In toto six different IF chain types have been recognized and characterized in terms of sequence features. Molecular assembly follows an increasingly well-understood pathway and results in four highly specific IF depending on chain type. Of these, trichocyte keratin intermediate filaments display two distinct structures. The first is assembled at the base of the hair follicle in living cells and is produced in a reducing environment. Structurally, it is equivalent to the IF formed in epidermal keratin. The second form occurs after cell death and is found at a point about half-way along the hair follicle. The observed structural changes involve both axial and lateral re-organisation of the subfibrillar elements and occur as a direct consequence of the chemical changes that arise as the environment changes from a reducing to an oxidizing one. In contrast, other intermediate filaments such as vimentin and neurofilaments have but a single structural form noting, however, that various chain combinations copolymerise in vivo. This endows the IFs with unique properties even though they are based on a structural framework that is common to each (but distinct from that in the keratins). Lamin IF represent the fourth structural form observed.

Introduction

Intermediate filaments (IF) represent a family of closely-related, nonpolar cytoskeletal filaments with diameters that typically lie in the range 10 ± 2 nm (for reviews of various aspects of IF structure and function, see refs. 1-7). As such and as their name suggests, they are intermediate in size between two other major classes of polar filamentous structures that had previously been studied and characterized in eukaryotic cells—the microtubules and the microfilaments. The IF are composed of molecules that originate from a wide variety of cells. Furthermore, they are expressed in a complex manner that is controlled by tissue-specific developmental programmes of differentiation.[2,8-11] Amongst the IF proteins the keratins are found in trichocyte and epithelial cells, vimentin in cells of mesenchymal origin, desmin in myogenic cells, glial fibrillary acidic protein in glial cells and astrocytes, peripherin and neurofilament proteins in neurons, nestin in

*Corresponding Author: David A.D. Parry—Institute of Fundamental Sciences, Massey University, Private Bag 11-222, Palmerston North, New Zealand. Email: d.parry@massey.ac.nz

Fibrous Proteins, edited by Thomas Scheibel. ©2008 Landes Bioscience.

neuroepithelial cells and lamin proteins in the nuclear envelope. Irrespective of cell type origin, however, the IF as a group play a major structural/mechanical role by providing cells with sufficient resilience to resist both the severe mechanical and nonmechanical stresses to which they are normally exposed in vivo, thereby permitting the cells to function in the manner required. Therefore, IF can be considered as the sensors and integrators of the cellular interior under stress.[12] Nonetheless, it has become increasingly apparent that the IF are very much more than just a family of static structures with a mechanical role. Over the past 10 years or so evidence has emerged of their dynamic nature and of their functional attributes imparted by interactions with a variety of signalling, structural, chaperone and regulatory proteins.[2,3,6,13-18] Equally fascinating is the number of human diseases that have been shown to arise either directly or indirectly from mutations in IF proteins. Amongst these are those that occur in epidermal keratin proteins and which lead to numerous skin pathologies, such as epidermolysis bullosa simplex and epidermolytic hyperkeratosis,[19-21] those that occur in desmin and which result in muscle weakness and disorganization[22-24] and those that occur in the lamin proteins and which lead to muscular dystrophies, cardiomyopathies, progeria and a variety of other serious conditions.[25-28] From all of these points it is evident that IF are involved in many aspects of cellular function and structure and it is only by studying them in detail that a thorough appreciation can be gained of cellular features as diverse as apoptosis, cell division, disease mechanisms, chromatin organization and cell architecture. The importance of IF as a major constituent of eukaryotic cells is thus self-evident and well worth the effort that has been devoted to them over the past 25-30 years.

Characteristic Features of Keratin Sequences and Chain Typing

The amino acid sequences of all keratin IF chains contain a central rod domain region about 310-315 residues long that is conducive to the formation of an α-helical structure. This is recognizable from the α-helix-favouring amino acid composition of this region but, more importantly, from the presence of an underlying heptad substructure of the form $(a\text{-}b\text{-}c\text{-}d\text{-}e\text{-}f\text{-}g)_n$, where positions *a* and *d* are most frequently apolar residues such as leucine, isoleucine and valine.[29,30] On the basis of well-defined heptad repeats the rod domain sequence can be further subdivided into four coiled coil segments 1A, 1B, 2A and 2B (35, 101, 19 and 121 residues long respectively) and three (generally) nonhelical linkers L1, L12 and L2 (typically 10-14, 16 or 17 and 8 residues long respectively) (Fig. 1A). At the centre of segment 2B there is a "stutter" in the sequence, which is equivalent to a three-residue deletion from what is an otherwise continuous heptad repeat.[31] Segments 1B and 2B both display a regular linear disposition of acidic and basic residues[32] and in human keratins these have periods of 9.51 and 9.87 residues in segments 1B and 2B respectively.[33] This specifies various modes of molecular aggregation (section 1.3) through the formation of ionic interactions between molecules that are appropriately aligned axially. A further characteristic of IF sequences is the highly conserved sequences found at both ends of the rod domain i.e., at the N-terminal end of segment 1A and the C-terminal end of segment 2B. These are known respectively as the helix initiation and the helix termination motifs.

On the basis of sequence homology within the rod domain all keratin sequences may be classified as being either Type I (net acidic, pI values 4.5-5.5 and molecular weights 40-64 kDa) or Type II (neutral-basic, pI values 6.5-7.5 and molecular weights 52-68 kDa). Sequence differences have been observed, however, between the chains of trichocyte and epithelial keratins. Consequently, those in the former group (trichocyte keratins) have been designated Type Ia and IIa and those in the latter (epithelial keratins) Type Ib and IIb.[34] It has been reported[33] that the average identity between the Type Ia chains, the Type IIa chains, the Type Ib chains and the Type IIb chains are all very high (74, 80, 58 and 68% respectively) as is that between Type I chains as a whole and Type II chains as a whole (57 and 65% respectively).

The regions that terminate the rod domain are referred to as the head and tail domains. Sequence substructure with bilateral symmetry has been reported within each of them. In the case of the epidermal keratins E1, V1 and H1 subdomains form the N-terminal head and the H2, V2 and E2 subdomains form the C-terminal tail.[35] The H1 and H2 regions are so named as they are

Homologous regions, each with a high degree of sequence conservation across all of the epidermal keratins. H1 is 7 and 36 residue long for Type I and II chains respectively and H2 is zero and 20 residues long for Type II chains. In contrast to the conserved H regions, the V1 and V2 subdomains are rich in glycine and serine residues and, with a few important exceptions, are devoid of conserved features (thus Variable). The E1 and E2 subdomains (End) are located at the terminal ends of the chains and are both short and relatively basic. For the trichocyte keratins the head and tail domains display some similar features (the H subdomains) but they lack the equivalent of the V and E subdomains.[36] The hair keratins have a high content of cysteine residues that oxidize in vivo to form a network of disulphide bonds that are both intramolecular and intermolecular. The latter can form between IF proteins as well as between IF proteins and the intermediate filament-associated proteins (IFAP) that surround the IF and which form a matrix structure.[37-39]

Chain Assembly and Keratin Molecule Formation

There is strong evidence that keratin molecules contain two chains and that both a Type I and a Type II chain are required, thereby resulting in an obligate heterodimer structure.[40-42] Crosslinking, biochemical data and theoretical modelling have confirmed that the two chains forming the IF heterodimer are in axial register and that they are parallel to one another.[43-45] The molecule thus has a polar, tripartite structure with a coiled-coil rod domain about 45.5 nm long. This figure is derived from the approximate number of residues in the rod domain (310-315) and the average unit rise per residue in a coiled-coil conformation (0.1485 nm).

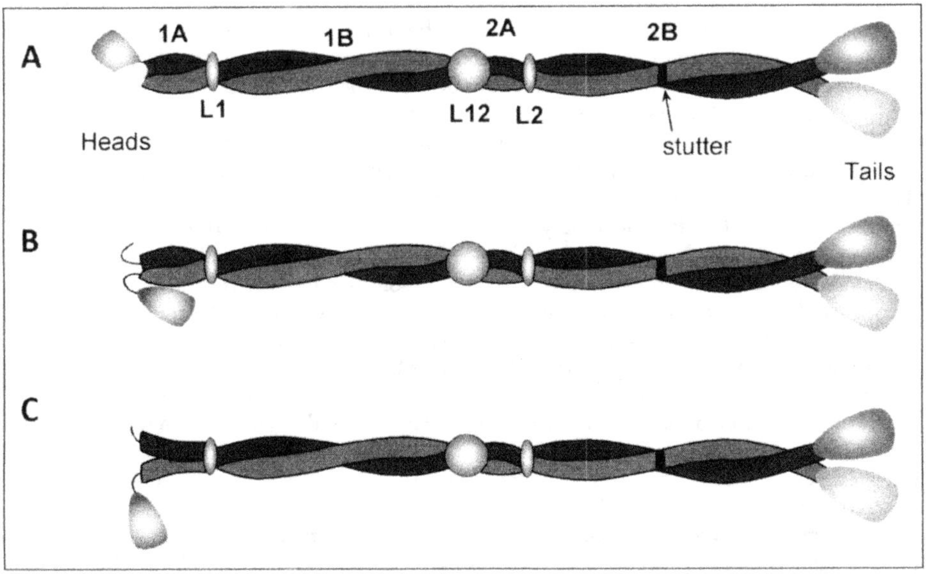

Figure 1. A) Model of an IF molecule illustrating a central rod domain and terminating head and tail domains. Four coiled-coil segments (1A, 1B, 2A, 2B) and three linkers (L1, L12, L2) comprise the main structural features of the rod domain. Other points of structural interest include a stutter near the centre of segment 2B, regularity in the linear disposition of the acidic and basic residues in segments 1B and 2 (i.e., 2A + L2 + 2B) and highly conserved helix initiation and termination motifs sited near the N-terminal end of segment 1A and the C-terminal end of segment 2B respectively, (B) the head domains are shown folded back over segment 1A to stabilize the coiled-coil structure and (C) the heads are shown distant from segment 1A thereby destabilising the structure and allowing the individual strands to separate. Figure from reference 7 and reprinted with the permission of Academic Press.

A point of some interest currently is the observation that segment 2A and linker L2 have a hendecad substructure (11 residue quasi-repeat) and that segment 2A alone has a heptad (7 residue) repeat.[46] What is the significance and meaning of this observation? The heptad substructure, first predicted more than 50 years ago by Crick[47] and subsequently recognized in the sequences of many fibrous and globular proteins, has 3.5 res/turn on average. As this is less than the average number of residues per turn in a standard α-helix (3.63), this implies a coiled-coil structure[47,48] in which right-handed α-helical chains will wind around one another in a left-handed manner (with pitch lengths now known to be typically 14-17 nm for a two-stranded structure). This type of assembly internalises the apolar residues and stabilises the conformation adopted through close packing of these residues. A variant of the heptad repeat—a hendecad substructure of the form $(a\text{-}b\text{-}c\text{-}d\text{-}e\text{-}f\text{-}g\text{-}h\text{-}i\text{-}j\text{-}k)_n$—also has apolar residues in key positions (*a*, *d*, *e* and *h*) akin to those in the heptad repeat. In this case there are 11 residues in 3 turns of α-helix (3.67 res/turn on average). As 3.67 is slightly greater than the average number of residues per turn in a standard α-helix (3.63) the right-handed α-helical chains will wind around one another in a right-handed manner with a long pitch length (say 80-100 nm). The message from these observations is that while segments 1A, 1B and 2B will form left-handed coiled-coils (a point confirmed from X-ray crystallographic studies of IF coiled-coil fragments) segment 2A and L2 may well form a short piece of right-handed coiled-coil or at least a region in which the two component α-helices lie almost parallel to one another. Experimental evidence, of course, will be required to confirm this prediction.

Unlike the rod domain the structures of both the head and tail domains are currently ill-defined but some features have been identified in the epidermal and the trichocyte keratins that will be pertinent. In the former the large V subdomains contain a number of large apolar or aromatic residues periodically but not regularly distributed amongst the glycine and serine residues. It has been proposed and evidence evinced that the V subdomains in both the head and tail domains will adopt an irregular, highly convoluted, flexible and dynamic structure—the glycine loop—and that this will be anchored through interactions between the large apolar and aromatic residues. NMR and theoretical analyses have provided support for this concept.[49-53]

In the case of trichocyte keratin terminal domains a number of observations have been made. The tail domains, for example, show a number of unique sequence characteristics.[36] In Type Ia chains there is an eight-fold contiguous but imperfect repeat of the quasi-triplet P-C-X. This sequence motif is indicative of a left-handed poly-proline II type of structure with three residues per turn, thereby positioning the cysteine residues along one edge where they would be capable of forming disulphide bonds with a similar region in a second molecule. In Type IIa chains the tail region immediately adjacent to the rod domain was shown to have a four-stranded β-sheet conformation with one face almost totally apolar and the other rich in cysteine and charged residues. It can easily be envisaged that the β-sheet regions from different molecules could interact with one another and so stabilize the internal structure of the IF. The Type Ia head domain sequences, however, reveal little structural information but, in contrast, the Type IIa head domains contain four contiguous nonapeptide quasi-repeats (G-G-F-G-Y-R-S-X-G) over the middle third of the sequence.[54] This is suggestive of a β-sheet conformation. Once again, the possibility arises that this could form an interaction with a similar region in a second IF molecule.

Although the heterodimeric conformation of the keratin molecule is the norm it has been proposed that it may be promiscuous (i.e., one of several Type I chains could form a heterodimer with one of several Type II chains) and that homodimers, albeit in small quantities and temporally transient, can exist, especially in embryonic tissues.[33]

It has been proposed[54] that the two-stranded coiled coil of segment 1A relies on the head domain folding back over it to stabilize the structure (Fig. 1B). Furthermore, it was suggested that if the head domain moved away the coiled coil would necessarily become less stable and that the two α-helical strands would separate (Fig. 1C), thereby providing the heads with a greater capacity to interact with either other IF molecules or IFAPs. These ideas relied in part on the observations that a dimer was formed only in the presence of the head domain[55] and that the crystal structure of segment 1A was not that of a two-stranded coiled coil but was, in fact, characterized by individual coiled α-helical strands.[56]

Molecular Assembly, Keratin IF Formation and Structure

The first detailed information on the structure of trichocyte keratin IF was derived from X-ray diffraction studies on porcupine quill. The IF in this material were more closely aligned (axially) than those found in either other "hard" keratins (such as hair) or indeed in the epidermal keratins in general. Porcupine quill thus allowed quantitative X-ray diffraction data to be collected and analysed (see, for example, refs. 57-59). The conclusions, in part, were that the keratin IF molecules had a high content of α-helical coiled-coil structure and that the molecules were aligned approximately parallel to the axis of the IF. The coherent length of the coiled coil was only about 5 nm, however, indicating that breaks occurred on average at about these distances.[58] Two possible surface lattice structures for oxidized keratin IF were determined by X-ray diffraction (the (8 + 0) and (7 + 1) models with the latter currently favoured[60,61] but it has not proved possible to achieve similar success for reduced keratin IF from any source. Nonetheless, a (8 + 0) surface lattice structure for reduced keratin IF has been deduced combining crosslinking and cryo-electron microscopy data.[62,63] STEM studies on trichocyte keratins have provided quantitative data on the mass per unit length of the IF[62,64] and these have indicated that each IF has about 32 chains in cross-section though possible polymorphism based on an eight-chain unit has been recognised. It follows from the STEM data that each surface lattice point is associated with a four-chain unit i.e., a pair of antiparallel molecules.[2]

The linear regularity of the acidic and basic residues in the two major lengths of α-helical coiled-coil in the rod domain was strongly suggestive that intermolecular ionic interactions were instrumental in driving the assembly of IF molecules into intact IF. On the basis of this premise it was proposed that three modes of alignment were likely to occur between antiparallel molecules.[30] In the first the 1B segments were largely overlapped (subsequently termed the A_{11} mode), in the second the 2B segments were largely overlapped (the A_{22} mode) and in the third the molecules as a whole were largely overlapped (the A_{12} mode). In all three cases the alignments were based on the maximization of intermolecular interactions between two 1B or two 2B segments (or a combination of both). Some evidence to support the A_{11} mode came from work on epidermal keratin[43] and wool[65,66]but definitive work came only from the use of DST crosslinking between spatially adjacent lysine residues in epidermal keratins K1/K10 and K5/K14.[44,45] Many crosslinks were identified and characterized and in every instance the data were compatible with one of the A_{11}, A_{22} and A_{12} modes of alignment (Fig. 2A,B,C). A few intramolecular crosslinks were also found, however, and these were consistent with the two chains in the molecule being parallel to one another and in axial register. Using the crosslink data as a whole it was possible to do a least squares analysis to determine the axially projected lengths of the linker regions (L1, L12 and L2) and the precise values of A_{11}, A_{22} and A_{12}. These are listed in Table 1. A fourth mode of molecular alignment (A_{CN}) involving a short head-to-tail overlap between similarly oriented (parallel) molecules arises directly, in effect, from the consecutive application of the A_{11} and A_{22} modes (Fig. 2D).

Equivalent data for hair keratin came some years later but were to provide a major surprise to those in the field.[67] Its origin lay in the basic cell biology of hair. Firstly, hair keratin IF are formed at the base of the hair follicle in the column of cells immediately above the dermal papilla. The intracellular environment is naturally a reducing one. Cell death occurs at a point further up the hair follicle and the environment changes to become an oxidizing one. Crosslinks were determined for both cases and although the modes of alignment were identical the actual values of the A_{11} stagger differed between the reducing and oxidizing states. Lateral changes in the packing of the molecules and in the subfibrillar elements present also occur.[68] Thus, hair keratin IF undergo a dynamic transition from a structure with A_{11}, A_{22} and A_{12} parameters that are closely akin to those in epidermal keratin IF to one in which the A_{22} and A_{12} parameters are maintained but where the A_{11} value is significantly altered. The transition results in many of the cysteine residues becoming axially aligned and hence capable of forming covalent disulphide bonds with other IF molecules.

Table 1. ***Structural parameters of IF derived from crosslinking studies***

	K1/5/10/14	Reduced α-keratin		Oxidised α-keratin			Vimentin	α-Internexin	Type III, Type IV and Type III/IV Copolymers
Fixed		*L1, L12, L2*	*L2, z_a/z_b*	*L1, L12, L2*		*z_a, z_b*			
z_a	79.74	78.15	74.48	50.00	47.14	50.03	23.89	-	25.06
z_b	110.71	112.18	111.78	133.30	130.71	133.33	139.53	137.00	136.43
Repeat	301.16	302.51	298.05	316.61	308.48	316.63	302.95	-	297.93
Overlap	7.03	5.67	2.42	-8.43	-11.72	-13.67	4.31	10.00	6.99
A_{12}	-0.70	3.30	-0.34	6.98	1.55	2.56	5.50	7.00	4.19
A_{11}	-111.41	-108.88	-112.12	-126.32	-129.16	-130.77	-134.03	-130.00	-132.24
A_{22}	189.75	193.63	185.93	190.28	179.33	185.86	168.92	-	165.68
L1	14.84	14.84	8.96	14.84	7.21	9.56	11.47	14.50	10.99
L12	12.57	12.57	10.03	12.57	9.43	11.85	11.99	-	8.99
L2	4.77	4.77	5.52	4.77	4.11	5.52	7.79	9.00	8.94
Δz(1BU,1BD)	-3.91	-1.38	-0.54	-18.82	-17.85	-17.98	-20.72	-	-20.30
Δz(2BU,2BD)	2.56	6.45	6.46	3.10	3.57	3.91	-17.34	-19.00	-18.24
Δz(1BU, 2D)	-50.54 -	46.54	-44.31	-42.86	-40.74 -	42.02	-40.97	-42.50	-41.80

These strengthen the IF structure as a whole and ultimately endow hair with many of its special mechanical attributes. An interesting effect (or is it the cause?) of the reduced to oxidized transition in hair keratin[33,67] is that the head and tail domains become much more closely organized in the final form than that which occurs in the initial structure[69] (Fig. 3).

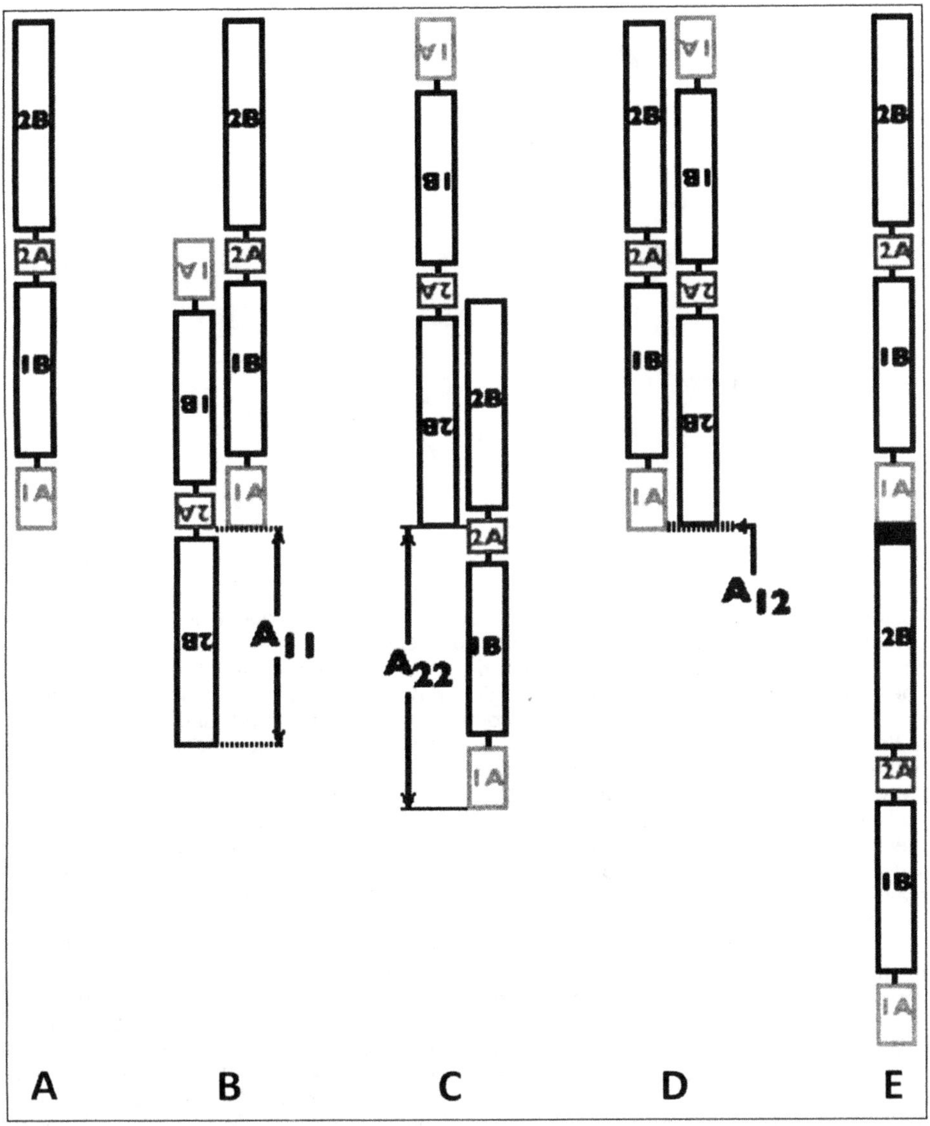

Figure 2. Schematic IF molecule (A) displaying four modes of assembly that are common in all IF. Three of these are unique and involve antiparallel molecules with (B) their 1B segments largely overlapped (A_{11}), (C) their 2B segments largely overlapped (A_{22}) and (D) their rod domains largely overlapped (A_{12}). (E) The fourth mode (A_{CN}) involves a short head-to-tail overlap between similarly directed molecules and is derived from the successive applications of the A_{11} and A_{22} modes.

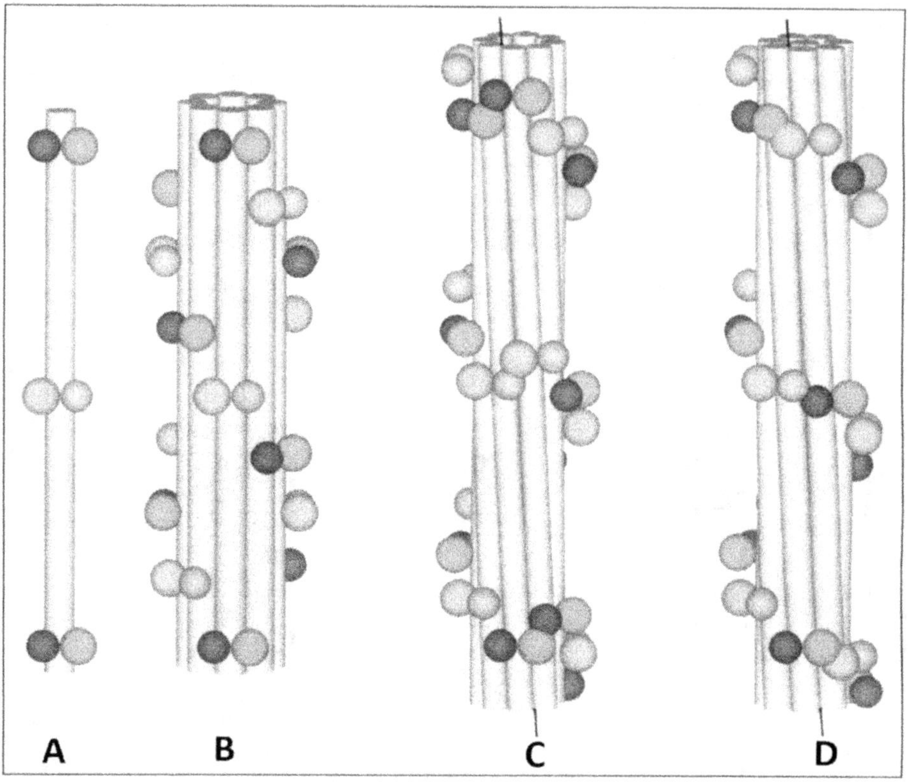

Figure 3. Generic models for the three-dimensional structure of reduced and oxidised trichocyte IF.[69] The IF can be regarded as composed of eight protofilaments each consisting of a pair of oppositely directed molecular strands. A) A single protofilament in the reduced IF. The yellow cylinders represent the helical domains of the constituent molecules, the green spheres represent the N-terminal domains of the Up molecular strands and the blue spheres to those of the Down molecular strands. The red spheres represent the C-terminal domains in the Up molecular strands and the orange spheres represent the C-terminal domains of the Down molecular strands. The axial period is 44.92 nm. B) The reduced IF is visualized as containing eight protofilaments arranged regularly on a ring of radius 3.5 nm. Each protofilament is related to its predecessor by a rotation of 45° and an axial displacement of 16.85 nm. The end groups fall on a two-start left-hand helix that, at low resolution, would appear as a diagonal banding pattern with a spacing of 22 nm. C) The (7 + 1) arrangement for oxidized IF in hydrated keratin is based on one of the two possible surface lattice structures derived from X-ray diffraction studies that are also compatible with other data. It differs from that of the (8 + 0) model for reduced keratin IF in several respects. Firstly, seven protofilaments are assembled on a ring of radius 3.0 nm, each related to its predecessor by a rotation of 30.7° and an axial displacement of 19.82 nm. These surround a central, straight protofilament. Secondly, it follows that the surface lattice structure is different from that in the reduced model but the axial period in an individual protofilament remains unchanged at 47.0 nm. Thirdly, the end domains are now in much closer proximity than in the reduced model and give the appearance of being distributed on a helix of pitch 23.5 nm. The transition from (A) the reduced to (B) the oxidized form of trichocyte keratin results in a major reorganization of the terminal head and tail domains on the surface of the IF. In the oxidized case the terminal domains are much closer to one another than they are in the reduced form[69] and it has been suggested that terminal domain interactions might provide the driving force for the structural reorganization that takes place on oxidization. Figure from ref 69 and reprinted with permission of Elsevier Press. A color version of this image is available online at www.eurekah.com.

Terminology commonly used in the field defines a molecular strand as a filament containing similarly directed molecules linked by a short head-to-tail overlap or gap in many cases (A_{CN}). The molecular strand is thus one molecule or two chains wide (excluding the gap/overlap regions; Fig. 4B). Pairs of antiparallel molecular strands defined by A_{11} and A_{22} interactions are termed protofilaments (two molecules or four chains in width; Fig. 4C). In turn, pairs of protofilaments linked by A_{12} interactions define a protofibril (four molecules or eight chains in width; Fig. 4D). In all cases, however, these subfibrillar elements are not intermediates in the assembly process but have proved useful descriptors of specific post-assembly aggregates. In particular it is useful to consider the reduced form of the trichocyte keratin IF in terms of eight protofilaments (or four protofibrils) that are structurally equivalent and arranged on an annulus of radius 3.5 nm, thus giving rise to a hollow core structure as observed (the 8 + 0 model). In contrast, in the oxidized form the peak in the electron density is found at a radius of 3.0 nm, thereby indicating radial compaction between the formation of the IF in a reducing medium and oxidation that occurs on cell death.[68] In addition the structure of the IF is best fitted to a (7 + 1) model with a central protofilament and seven others arranged on an annulus as noted above. In this case the protofibrils are structurally non-equivalent.

Although the protofilament and protofibrillar sub-fibrillar elements are useful concepts post-IF formation the filaments are actually formed by a mechanism that involves (a) the rapid lateral aggregation of about eight tetramers. Each of these consists of a half-staggered antiparallel pair of molecules assembled in the A_{11} mode and the structure thus formed—a unit-length-filament (ULF)—is about 55 nm long. Assembly at the two-to-four molecule level in epidermal keratin is facilitated through interactions involving the H1 zones in the head domain of Type II chains.[44,70,71] Since the equivalent sequences are present in the trichocyte keratins the same role in assembly as that shown experimentally for the epidermal keratins is highly probable; (b) an elongation step involving the axial aggregation of ULF to generate immature IF about 16 nm in diameter and many μm long. The dimers within this structure are believed to undergo reorganization during this process; (c) a radial compaction step occurs which results in close packing of the molecular filaments and the formation of IF with diameters of 8-12 nm.[12,67,72-74]

Comparison with Other Intermediate Filaments

With regard to cell type the sequences of the expressed IF proteins display characteristic differences and marked similarities. An understanding of the role that these subtle but important sequence differences play in molecular structure and assembly and ultimately in function is clearly of some importance and Section 2 will provide a description of the current state of knowledge.

Sequence Similarities/Differences and Chain Typing

Comparison of IF sequences, of which several hundred have now been determined (see www.interfil.org for human IF sequences), confirm most of the features recorded earlier for the keratins i.e., the presence of a central rod domain with four coiled-coil segments, conserved helix initiation and termination motifs, a stutter close to the centre of segment 2B and a well-defined period in the linear distribution of the acidic and the basic residues in segments 1B and 2B. A key observation, however, lies with the recognition that the rod domain sequences in the Type I and Type II keratins are clearly different from those in all other IF chains. On the basis of sequence homology a number of different chain types have now been proposed. These include the Type III chains (vimentin, desmin, glial fibrillary acidic protein, peripherin and syncoilin), the Type IV chains (neurofilament proteins NF-L, NF-M, NF-H and α-internexin) and the Type V chains (nuclear lamin proteins). Nestin and synemin have been designated Type VI IF chains. Consensus sequences for the rod domain region of Types I-V IF chains have been compiled.[75]

The head and tail domains display major and very characteristic sequence differences based on chain type. These include the following: (a) in a wide range of species the head domain in Types III and IV IF chains contain a conserved nonapeptide of the form S-S-Y-R-R-T-F-G-G. It has been shown that this is important for regular assembly to proceed;[76] (b) the tail domain of the Type IV

Figure 4. Possible sub-fibrillar elements in an intermediate filament that are larger than an IF molecule. These include (A) a molecular strand consisting of similarly directed molecules interacting via head-to-tail interactions (A_{CN}), (B) a protofilament consisting of a pair of antiparallel molecular strands interacting through A_{11} and A_{22} interactions and (C) a protofibril consisting of a pair of antiparallel protofilaments linked through A_{12} interactions.

IF chains can be subdivided into several clearly defined regions that are E-rich (the E-segment), conserved in K and E (KE segment), contain frequent KSP repeats (KSP segment) and, finally, a KEP region at the C-terminus (KEP segment). The KSP repeats represent multiple sites of phosphorylation and are thus of functional significance in determining such features as axonal diameter, conduction velocity, transport and associations with other components in the cytoskeleton (see ref 2 for a summary of earlier work); (c) the head domains of Type VI nestin contain quasi-repeats 11 residues long based on exact repeats of 22 residues in human and 44 residues in hamster and rat.[3,77] The structural implication of these repeats are not known; (d) the CaaX box at the C-terminal end of the Type V lamin chain that allows isoprenylation to proceed, thus targetting the lamin chains to the nuclear envelope. There are, of course, many other features of importance in the head and tail domain sequences. Indeed, many data are now available[7] but space limitations have precluded all but a very brief indication of some of the most interesting features that have been observed thus far. Disappointingly, crystal structures of head and tail domains remain elusive and only the tail domain of lamin has thus far been solved at atomic resolution.[78,79]

Chain Assembly and Intermediate Filament Molecule Formation

As summarized by Parry[7] some IF chains can homopolymerise (vimentin, desmin, peripherin, glial fibrillary acidic protein, NF-L and nestin) whilst others can heteropolymerise in the presence of a suitable partner. The latter include nestin (Type VI) with vimentin or α-internexin (both Type III), synemin (Type VI) with desmin or vimentin (both Type III), NF-M or NF-H (both Type IV) with NF-L (Type IV), as well as keratin (Type I) with keratin (Type II) as previously discussed. At this time, however, there is no evidence that any keratin chain (Type I or Type II) or lamin chain (Type V) is able to copolymerise with any Type III or Type IV IF chain.

There is compelling biochemical and crosslinking evidence that each IF molecule, irrespective of its source, contains a pair of chains. Without exception the chains are similarly directed and lie in axial register (see, for example, the reviews in refs 1-5). The length of the rod domain in IF (excluding lamin) can be estimated from the known axial rise per residue in a coiled-coil conformation (0.1485 nm) and the axial lengths of the linkers determined from crosslinking studies. Typically, the lengths of the rod domain determined for all IF molecules lie in the range 45.5 ± 0.3 nm.

Lamin, which is comprised of Type V chains, has a number of structural characteristics that distinguish them from the remainder of the family of IF chains. Firstly, segment 1B contains an additional 42 residues (six heptads) relative to other IF chains[80] though there is some debate as to whether a single insert of this size occurs or whether two inserts exist with a total length of 42 residues. Irrespective of that point, however, the length of segment 1B will be longer than in other IF chains by about 6.2 nm. Secondly, linker L1 in lamin is predicted to be α-helical, thereby implying that segments 1A and 1B effectively form a continuous structure. However, a stutter of the type first recognized in segment 2B[31] is also predicted to interrupt the phasing of the underlying heptad repeat at the site of the former linker L1.[81] This will lead to a short region in the molecule where the coiled-coil will unwind a little, as has been characterized crystallographically in segment 2B in vimentin.[82] Thirdly, linker L12 too is predicted to be largely α-helical and since linker L2 has been described earlier in terms of an underlying hendecad repeat[46] it seems that the rod domain of the lamin chains will be virtually entirely α-helical. It is therefore likely to adopt a coiled-coil structure over almost its entire length (mainly left-handed but with a short region possibly right-handed in the vicinity of what was earlier designated segment 2A and linker L2).

Molecular Assembly, IF Formation and Structure

The assembly of Types III, IV and VI IF molecules to form intact IF uses the same modes of aggregation as those described in Section 1.3 for the keratins (A_{11}, A_{22}, A_{12} and A_{CN}). An important difference, however, lies in the observation that the values of the axial staggers are not the same (Table 1) for A_{11} and A_{22} (about -132 and 166 coiled-coil residue equivalents compared with -110 and 191 respectively for the epidermal/reduced keratins). They are similar, however, for the A_{12} mode (about 3-4 coiled-coil residue equivalents on average in both cases).[77,83,84] Equivalent crosslinking studies have not been performed for cytoplasmic invertebrate IF proteins or for Type V lamins but electron microscope and other observations have indicated a different mode of assembly dictated initially by the head-to-tail assembly of molecules to generate a beaded string-like structure[85] The IF as a whole thus comprise a family of four unique but highly related structures—(a) epidermal and reduced keratin IF (Type I and Type II chains), (b) oxidized keratin IF (Type I and Type II chains), (c) those typified by vimentin and neurofilament IF (which involve various combinations of Types III, IV and VI chains) and (d) nuclear lamin IF (Type V chains).

Once again, STEM studies have been utilized to assess the numbers of chains in the IF cross-section. In the case of desmin the mass/unit length is about 38 kDa/nm^{-1}. This corresponds to about 32 chains,[4] a value that is directly comparable to that seen in trichocyte keratin. The step-wise assembly of desmin molecules follows the same path as already shown for the keratins (Section 1.3) in that it too involves (a) a rapid lateral assembly of tetramers to form unit-length filaments (ULF), (b) reorganization of the molecules within the ULF prior to longitudinal annealing, thereby forming long but loosely packed IF-like assemblies and (c) radial compaction to form mature IF.[73] It would be surprising if the major features of this generalised model were not followed closely in all IF excluding, for the reasons noted previously, those formed in the nuclear lamina.

Polymorphism, however, remains an interesting phenomena in IF as illustrated with vimentin IF. In this case the mass/unit length is about 56 kDa/nm^{-1} and this leads to about 48 chains in cross-section.[4,86] STEM data for IF as a whole display peaks in the mass/unit length profiles at the equivalent of about 16, 24, 32, 40 and 48 chains in section (see a summary of the data in refs 2 and 3). There is also evidence that the mass/unit length varies along the length of a single IF.[4] In all cases an eight-chain subunit is evident thus providing support for the existence of a discrete sub-IF structure—the protofibril. Nonetheless, polymorphism remains a phenomenon that is difficult to understand. For example, recent cryo-electron tomographic studies on vimentin[87] have revealed a four-stranded protofibrillar organization with a right-handed supertwist (pitch length 96 nm). In contrast, the STEM data had indicated a five-stranded protofibrillar structure whereas a four-stranded one has observed. The X-ray diffraction data for keratin IF also indicate a four-stranded structure. Generality has yet to be proven but a model with 32 chains in section for all IF is appealing. If this does prove to be the case then the STEM data might need to be re-analysed.

Summary

Some 40-50 years ago structural studies on the microfibrils in the hard α-keratins (later termed intermediate filaments) were largely confined to X-ray diffraction studies on quill and to allied biochemical investigations on wool. Importantly, these pioneering studies led by Fraser and Crewther at the CSIRO Division of Protein Chemistry in Australia, set the scene for new cell biology observations made in the late 1970s and early 1980s, which showed that filaments intermediate in size between the microtubules and the microfilaments were a ubiquitous component of eukaryotic cells. These "intermediate filaments" as they then became termed were shown to contain chains with sequences closely related to those of the keratins studied previously. The IF field thus effectively dates from these two sets of observations. Since then, of course, a great deal of new and exciting experimental data have been derived from a wide variety of techniques and, when allied to theoretical and modelling results, these have yielded an increasingly clear picture of the structure of the family of intermediate filaments and the role that each plays in vivo. It cannot be doubted that provided the current rate of progress can be maintained many of the uncertainties in the field will be resolved over the coming decade, thereby revealing the secrets that make the IF such an important and fundamental component of the cell.

References

1. Fuchs E, Weber K. Intermediate filaments; structure, dynamics, function and disease. Annu Rev Biochem 1994; 63:345-382.
2. Parry DAD, Steinert PM. Intermediate FilamentStructure. Heidelberg: Springer-Verlag, 1995.

3. Parry DAD, Steinert PM. Intermediate filaments: molecular architecture, assembly, dynamics and polymorphism. Quart Rev Biophys 1999; 32:99-187.
4. Herrmann H, Aebi U. Intermediate filaments and their associates: multi-talented structural elements specifying cytoarchitecture and cytodynamics. Curr Opin Cell Biol 2000; 12:79-90.
5. Herrmann H, Aebi U. Intermediate filaments: molecular structure, assembly mechanism and integration into functionally distinct intracellular scaffolds. Annu Rev Biochem 2004; 73:749-789.
6. Herrmann H, Hesse M, Reichenzeller M et al. Functional complexity of intermediate filament cytoskeletons: from structures to assembly to gene ablation. Int Rev Cytol 2003; 223:83-175.
7. Parry DAD. Microdissection of the sequence and structure of intermediate filament chains. Adv Pro Chem 2005; 70:113-142.
8. Lazarides E. Intermediate filaments as mechanical integrators of cellular space. Nature 1980; 293:249-256.
9. Lazarides E. Intermediate filaments: a chemically heterogeneous, developmentally regulated class of proteins. Annu Rev Biochem 1982; 51:219-250.
10. Herrmann H, Aebi U. Intermediate filament assembly: temperature sensitivity and polymorphism. Cell Mol Life Sci 1999; 55:1416-1431.
11. Langbein L, Schweizer J. Keratins of the human hair follicle. Int Rev Cytol 2005; 243:1-78.
12. Parry DAD, Strelkov SV, Burkhard P et al. Towards a molecular description of intermediate filament structure and assembly. Exp Cell Res 2007; 313:2204-2216.
13. Caulin C, Ware CF, Magin T et al. Keratin-dependent, epithelial resistance to tumor necrosis factor-induced apoptosis. J Cell Biol 2000; 149:17-22.
14. Gilbert S, Loranger A, Daigle N et al. Simple epithelium keratins 8 and 18 provide resistance to Fas-mediated apoptosis. The protection occurs through a receptor-targeting modulation. J Cell Biol 2001; 154:763-773.
15. Leung CL, Liem RKH, Parry DAD et al. The plakin family. J Cell Sci 2001; 114:3409-3410.
16. Magin TM, Vijayaraj P, Leube RE. Structural and regulatory functions of keratins. Exp Cell Res 2007; 313:2021-2032.
17. Ivaska J, Pallar H-M, Nevo J et al. Novel functions of vimentin in cell adhesion, migration and signaling. Exp Cell Res 2007; 313:2050-2062.
18. Sihag RK, Inagaki M, Yamaguchi T et al. Role of phosphorylation on the structural dynamics and function of types III and IV intermediate filaments. Exp Cell Res 2007; 313:2098-2109.
19. Fuchs E. Keratins and the skin. Annu Rev Cell Dev Biol 1995; 11:123-153.
20. DiGiovanna JJ, Bale SJ, Steinert PM. Epidermolytic hyperkeratosis. In: Jameson JL ed. Principles of Molecular Medicine. Totawa: Humana Press Inc., 1998:707-712.
21. Lane EB, McLean W. Keratins and skin disorders. J Pathol 2004; 204:355-366.
22. Bär H, Strelkov SV, Sjöberg G et al. The biology of desmin filaments: how do mutations affect their structure, assembly and organization? J Struct Biol 2004; 148:137-152.
23. Bär H, Mucke N, Ringler P et al. Impact of disease mutations on the desmin filament assembly process. J Mol Biol 2006; 360:1031-1042.
24. Paulin D, Huet A, Khanamyrian L et al. Desminopathies in muscle disease. J Pathol 2004; 204:418-427.
25. Bonne G, Di Barletta MR, Varnous S et al. Mutations in the gene encoding lamin A/C cause autosomal dominant Emery-Dreifuss muscular dystrophy. Nat Genet 1999; 21:285-288.
26. Burke B, Stewart CL. Life at the edge: the nuclear envelope and human disease. Nat Rev Mol Cell Biol 2002; 3:575-585.
27. Worman HJ, Courvalin JC. Nuclear envelope, nuclear lamina and inherited disease. Int Rev Cytol 2005; 246:231-279.
28. Capell BC, Collins FS. Human laminopathies: nuclei gone genetically awry. Nat Rev Genet 2006; 7:940-952.
29. Geisler N, Kaufmann E, Weber K. Protein-chemical characterization of three structurally distinct domains along the protofilament unit of 10 nm filaments. Cell 1982; 30:277-286.
30. Crewther WG, Dowling LM, Steinert PM et al. Structure of intermediate filaments. Int J Biol Macromol 1983; 5:267-274.
31. Brown JH, Cohen C, Parry DAD. Heptad breaks in α-helical coiled coils: stutters and stammers. Proteins: Struct Funct Genet 1996; 26:134-145.
32. Parry DAD, Crewther WG, Fraser RDB et al. Structure of α-keratin: structural implication of the amino acid sequences of the type I and type II chains segments. J Mol Biol 1977; 113:449-454.
33. Smith TA, Parry DAD. Sequence analyses of type I and type II chains in human hair and epithelial keratin intermediate filaments: promiscuous obligate heterodimers, type II template for molecule formation and a rationale for heterodimer formation. J Struct Biol 2007; 158:344-357.

34. Conway JF, Parry DAD. Intermediate filament structure. 3. Analysis of sequence homologies. Int J Biol Macromol 1988; 10:79-98.
35. Steinert PM, Parry DAD, Idler WW et al. Amino acid sequences of mouse and human epidermal type II keratins of M_r 67000 provide a systematic basis for the structural and functional diversity of the end domains of keratin intermediate filament subunits. J Biol Chem 1985; 260:7142-7149.
36. Parry DAD, North ACT. Hard α-keratin intermediate filament chains: substructure of the N- and C-terminal domains and the predicted structure and function of the C-terminal domains of the type I and type II chains. J Struct Biol 1998; 122:67-75.
37. Fraser RDB, MacRae TP, Conformation in Fibrous Proteins and Related Synthetic Polypeptides. New York and London: Academic Press, 1973.
38. Fraser RDB, MacRae TP, Sparrow LG et al. Disulphide bonding in α-keratin. Int J Biol Macromol 1988; 10:106-112.
39. Parry DAD, Smith TA, Rogers MA et al. Human hair keratin-associated proteins: sequence regularities and structural implications. J Struct Biol 2006; 155:361-369.
40. Steinert PM. The two-chain coiled-coil molecule of native epidermal keratin intermediate filaments is a type I—type II heterodimer. J Biol Chem 1990; 265:8766-8774.
41. Coulombe PA, Fuchs E. Elucidating the early stages of keratin filament assembly. J Cell Biol 1990; 11:153-169.
42. Herrling J, Sparrow LG. Interactions of intermediate filament proteins from wool. Int J Biol Macromol 1991; 13:115-119.
43. Parry DAD, Steven AC, Steinert PM. The coiled-coil molecules of intermediate filaments consist of two parallel chains in exact axial register. Biochem Biophys Res Commun 1985; 127:1012-1018.
44. Steinert PM, Marekov LN, Fraser RDB et al. Keratin intermediate filament structure: crosslinking studies yield quantitative information on molecular dimensions and mechanism of assembly. J Mol Biol 1993; 230:436-452.
45. Steinert PM, Marekov LN, Parry DAD. Conservation of the structure of keratin intermediate filaments: molecular mechanism by which different keratin molecules integrate into pre-existing keratin intermediate filaments during differentiation. Biochemistry 1993; 32:10046-10056.
46. Parry DAD. Hendecad repeat in segment 2A and linker L2 of intermediate filament chains implies the possibility of a right-handed coiled-coil structure. J Struct Biol 2006; 155:370-374.
47. Crick FHC. The packing of α-helices: simple coiled-coils. Acta Cryst 1953; 6:689-697.
48. Pauling L, Corey RB, Branson HR. The structure of proteins: two hydrogen-bonded helical configurations of the polypeptide chain. Proc Natl Acad Sci USA 1951; 37:205-211.
49. Mack JW, Torchia DA, Steinert PM. Solid-state NMR studies of the dynamics and structure of mouse keratin intermediate filaments. Biochemistry 1988; 27:5418-5426.
50. Conway JF, Fraser RDB, MacRae TP et al. Protein chains in wool and epidermal keratin IF: structural features and spatial arrangement. In: Rogers GE, Reis PJ, Ward KA, Marshall RC, eds. The Biology of Wool and Hair. London: Chapman and Hall, 1989:127-144.
51. Steinert PM, Torchia D, Mack JW. Structural features of keratin intermediate filaments. In: Rogers GE, Reis PJ, Ward KA, Marshall RC, eds. The Biology of Wool and Hair. London: Chapman and Hall, 1989:157-167.
52. Conway JF, Parry DAD. Structural and spatial organization of intermediate filament and nuclear lamin proteins. In: Aebi U Engel J eds Cytoskeletal and Extracellular Proteins: Structure Interactions and Assembly. Berlin, Springer-Verlag 1989; vol 3:140-149.
53. Steinert PM, Mack JW, Korge BP et al. Glycine loops in proteins: occurrence in certain intermediate filament chains, loricrins and single stranded RNA binding proteins. Int J Biol Macromol 1991; 13:130-139.
54. Parry DAD, Marekov LN, Steinert PM et al. A role for the 1A and L1 rod domain segments in head domain organization and function of intermediate filaments: structural analysis of trichocyte keratin. J Struct Biol 2002; 137:97-108.
55. Strelkov SV, Herrmann H, Geisler N et al. Divide-and-conquer crystallographic approach towards an atomic structure of intermediate filaments. J Mol Biol 2001; 306:773-781.
56. Strelkov SV, Herrmann H, Geisler N et al. Conserved segments 1A and 2B of the intermediate filament dimer: their atomic structures and role in assembly. EMBO J 2002; 21:1255-1266.
57. Fraser RDB, MacRae TP, Miller A. The coiled-coil model of α-keratin structure. J Mol Biol 1964; 10:147-156.
58. Fraser RDB, MacRae TP, Miller A. X-ray diffraction pattern of α-fibrous proteins. J Mol Biol 1965; 14:432-442.
59. Fraser RDB, MacRae TP, Suzuki E. Structure of the α-keratin microfibril. J Mol Biol 1976; 108:435-452.
60. Fraser RDB, MacRae TP. Intermediate filament structure. Biosci Rep 1985; 5:573-579.

61. Fraser RDB, MacRae TP. Surface lattice in α-keratin filaments. Int J Biol Macromol 1988; 10:178-184.
62. Watts NR, Jones LN, Cheng N. Cryo-electron microscopy of trichocyte (hard α-keratin) intermediate filaments reveals a low-density core. J Struct Biol 2002; 137:109-118.
63. Fraser RDB, Steinert PM, Parry DAD. Structural changes in trichocyte keratin intermediate filaments during keratinization. J Struct Biol 2003; 142:266-271.
64. Jones LN, Simon M, Watts NR et al. Intermediate filament structure: hard α-keratin. Biophys Chem 1997; 68:83-93.
65. Woods EF, Inglis A. Organization of the coiled-coils in the wool microfibril. Int J Biol Macromol 1984; 6:277-283.
66. Sparrow LG, Dowling LM, Loke VY et al. Amino acid sequence in wool keratin IF proteins. In: Rogers GE, Reis PJ, Ward KA, Marshall RC, eds. The Biology of Wool and Hair. London: Chapman and Hall, 1989:145-155.
67. Wang H, Parry DAD, Jones LN et al. In vitro assembly and structure of trichocyte keratin intermediate filaments: a novel role for stabilization by disulfide bonding. J Cell Biol 2000; 151:1459-1468.
68. Fraser RDB, Parry DAD. Structural changes in the trichocyte intermediate filaments accompanying the transition from the reduced to the oxidized form. J Struct Biol 2007; 159:36-45.
69. Fraser RDB, Parry DAD. The three-dimensional structure of trichocyte (hard α-) keratin intermediate filaments: features of the molecular packing deduced from the sites of induced crosslinks. J Struct Biol 2005; 151:171-181.
70. Steinert PM, Parry DAD. The conserved H1 domain of the type II keratin 1 chains plays an essential role in the alignment of nearest neighbor molecules in mouse and human keratin 1/keratin 10 intermediate filaments at the two- to four-molecule level of structure. J Biol Chem 1993; 268:2878-2887.
71. Hatzfeld M, Burba M. Function of type I and type II keratin head domains: their role in dimer, tetramer and filament formation. J Cell Sci 1994; 107:1959-1972.
72. Eichner R, Rew P, Engel A et al. Human epidermal keratin filaments: studies on their structure and assembly. Ann N Y Acad Sci 1985; 455:381-402.
73. Herrmann H, Häner M, Brettel M et al. Characterization of distinct early assembly units of different intermediate filament proteins. J Mol Biol 1999; 286:1403-1420.
74. Herrmann H, Wedig T, Porter RM et al. Characterization of early assembly intermediates of recombinant human keratins. J Struct Biol 2002; 137:82-96.
75. Smith TA, Strelkov SV, Burkhard P et al. Sequence comparisons of intermediate filament chains: evidence of a unique functional/structural role for the coiled-coil segment 1A and linker L1. J Struct Biol 2002; 137:128-145.
76. Herrmann H, Hofmann I, Franke WW. Identification of a nonapeptide motif in the vimentin head domain involved in intermediate filament assembly. J Mol Biol 1992; 223:637-650.
77. Steinert PM, Chou YH, Prahlad V et al. A high molecular weight intermediate filament-associated protein in BHK-21 cells is nestin, a type IV intermediate filament protein: limited co-assembly in vitro to form heteropolymers with type III vimentin and type IV α-internexin. J Biol Chem 1999; 274:9881-9890.
78. Dhe-Paganon S, Werner ED, Chi YI et al. Structure of the globular tail of nuclear lamin. J Biol Chem 2002; 277:17381-17384.
79. Krimm I, Ostlund C, Gilquin B et al. The Ig-like structure of the C-terminal domain of lamin A/C, mutated in muscular dystrophies, cardiomyopathy and partial lipodystrophy. Structure 2002; 10:811-823.
80. McKeon FD, Kirschner MW, Caput D. Homologies in both primary and secondary structure between nuclear envelope and intermediate filament proteins. Nature 1986; 319:463-468.
81. Parry DAD, Conway JF, Steinert PM. Structural studies on lamin: similarities and differences between lamin and intermediate-filament proteins. Biochem J 1986; 238:305-308.
82. Strelkov SV, Burkhard P. Analysis of α-helical coiled coils with the program TWISTER reveals a structural mechanism for stutter compensation. J Struct Biol 2002; 137:54-64.
83. Steinert PM, Marekov LN, Parry DAD. Diversity of intermediate filament structure. Evidence that the alignment of coiled-coil molecules in vimentin is different from that in keratin intermediate filaments. J Biol Chem 1993; 268:24916-24925.
84. Steinert PM, Marekov LN, Parry DAD. Molecular parameters of type IV α-internexin and type IV—type III α-internexin-vimentin copolymer intermediate filaments. J Biol Chem 1999; 274:1657-1666.
85. Geisler N, Schunemann K, Weber K et al. Assembly and architecture of invertebrate cytoplasmic intermediate filaments reconcile features of vertebrate cytoplasmic and nuclear lamin-type intermediate filaments. J Mol Biol 1998; 282:601-617.
86. Herrmann H, Häner M, Brettel M et al. Structure and assembly properties of the intermediate filament protein vimentin: the role of its head, rod and tail domains. J Mol Biol 1996; 264:933-953.
87. Goldie KN, Wedig T, Mitra AK et al. Dissecting the 3-D structure of vimentin intermediate filaments by cryo-electron tomography. J Struct Biol 2007; 158:378-385.

CHAPTER 7

Tubulin and the Dynamic, Multifunctional Microtubules It Forms

Kenneth H. Downing*

Introduction

Tubulin is a member of the class of proteins that are globular themselves but that form fibers in aggregate. It is the principal protein component of microtubules (MTs), which are hollow cylindrical tubes that carry out a variety of functions within the cell. MTs serve as tracks for motor molecules to carry cargo around within the cell and are also involved in maintaining cell shape. Many other proteins interact with MTs in addition to the motors and many of these proteins regulate the various functions. Because their functions are essential for completion of the cell cycle, disrupting the normal behavior of MTs has been a successful chemotherapeutic approach in cancer treatment. The importance of the roles of MTs as well as their fascinating physical and mechanical properties has made MTs a target in many areas of scientific study. These same properties and interactions with motor proteins have also made MTs promising functional elements in nanotechnology. We will review some of these properties and the protein structure that gives rise to them.

Roles and Structure of Microtubules

Microtubules are one of the main components of the cell cytoskeleton. They are frequently visualized in the fluorescence light microscope as a fibrous net that fills the cell during interphase (Fig. 1). During mitosis, this network is reorganized into the spindle that is involved in separation of the condensed chromosomes. Light microscopy of MTs both within cells and in vitro revealed that individual MTs frequently switch between phases of growth and shrinkage, a process termed "dynamic instability".[1] Figure 2 shows schematically the nature of MT length measurements made over a period of time, which can be seen both in vivo and in vitro. Growth tends to proceed at a steady rate until a "catastrophe", when the MT starts to shrink or depolymerize at a faster rate. At some stage "rescue" occurs and the MT starts to grow again. These events can be interspersed with "pauses" during which there is little change on length.

Microtubules in a cell often radiate from a particular region of the cell known as the "microtubule organizing center" (MTOC). The end that appears to be anchored in the MTOC is termed the "minus" end. Most of the dynamic instability is seen at the other end, the "plus" end. Occasionally MTs break due to excessive buckling as they grow within the confines of a cell. Similar instability is seen at the newly exposed minus end, although at significantly slower rates than at the plus end.

Microtubules are formed from the protein tubulin, which is a compact, globular protein. The basic subunit is a heterodimer containing alpha and beta tubulin monomers. Dimers are arranged head-to-tail in protofilaments and the protofilaments associate to form the wall of the MT (see Fig. 3). Each monomer is associated with a guanine nucleotide: alpha tubulin has GTP bound non-exchangeably, while GTP or GDP can bind exchangeably to beta tubulin. Polymerization of

*Kenneth H. Downing—Life Sciences Division, Lawrence Berkeley National Laboratory, Berkeley, California 94720, USA. Email: khdowning@lbl.gov

Fibrous Proteins, edited by Thomas Scheibel. ©2008 Landes Bioscience.

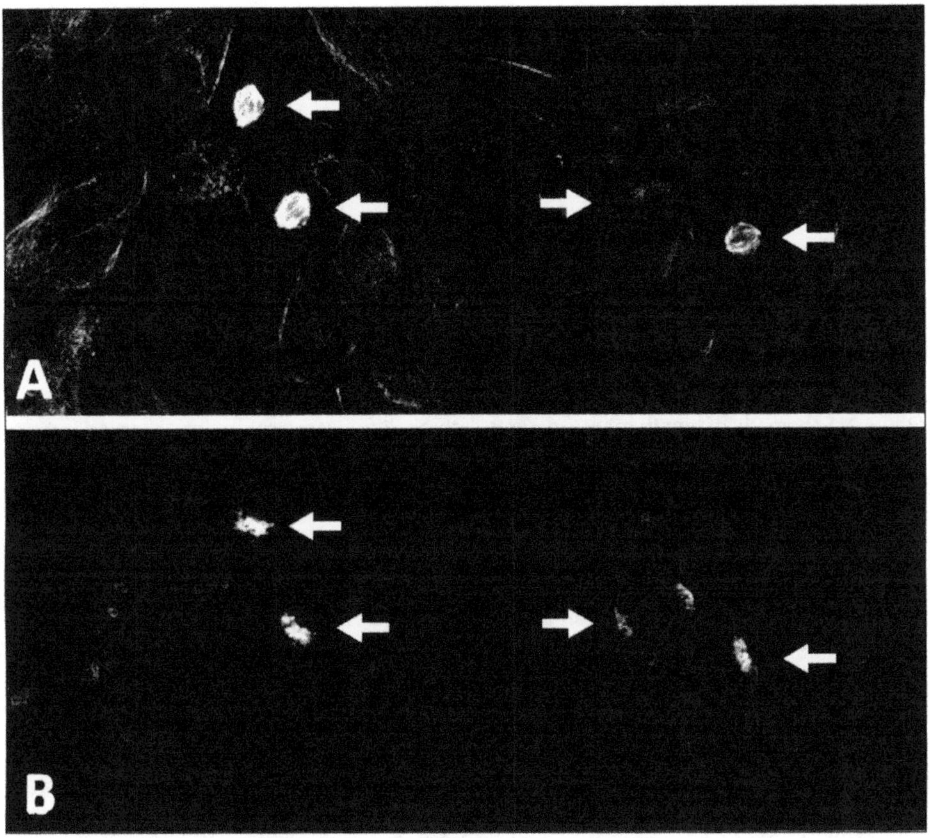

Figure 1. Fluorescence light micrograph a confluent mat of MDCK cells. The cells were fixed and the MTs immunostained with AlexaFluor 488 (Molecular Probes), visualized in (A), and nucleic acids stained with DAPI, visualized in (B). Most cells are seen to be essentially filled with an array of MTs, but several of the cells in this field are undergoing mitosis, where the MTs have reorganized to form the spindle. The right-pointing arrows mark spindles that are just beginning to separate chromosomes, which are mostly condensed into the metaphase plate. The left-pointing arrow marks a cell in which separation is nearly complete and MTs are beginning to repopulate the cell. Figure kindly provided by Alison Killilea and Haixin Sui.

the dimers into MTs generally requires the presence of GTP in beta and the GTP is hydrolyzed to GDP shortly after polymerization. The nucleotide in beta also becomes non-exchangeable upon polymerization. When MTs disassemble, the protofilaments sometimes separate from each other and curl into rings of about 30-40 nm diameter.[2,3] From this behavior it was concluded that protofilaments of GDP-tubulin preferred a curved configuration, but that in a MT lattice contacts between adjacent protofilaments keep the protofilaments straight. In the region near a growing end where the nucleotide has not yet been hydrolyzed, stronger contacts between protofilaments containing GTP-tubulin and/or the tendency of the GTP-containing protofilaments to remain straight, forms a "GTP cap" that prevents disassembly. Loss of the cap, which could happen by loss of GTP-containing dimers at the end, hydrolysis or exchange of their GTP, or addition of GDP-containing dimers (or a combination of these) leads to separation of protofilaments and the rapid shortening seen in catastrophes.

Among the many proteins that bind to MTs, some help to stabilize MTs and some to destabilize them. The level of dynamic behavior varies during the cell cycle in a highly organized way

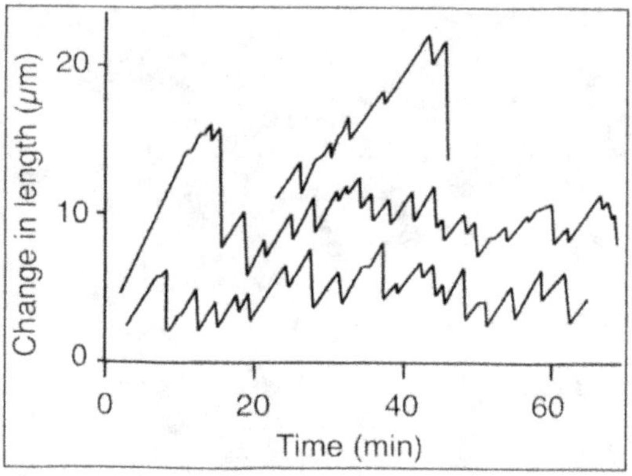

Figure 2. Schematic of length measurements of three different microtubules observed in vitro, showing periods of steady growth interspersed with shorter periods of rapid shrinkage. The average growth rate is about 2 µm/min, shrinkage about 50 µm/min. Note that the tie resolution is on the order of seconds in these measurements. Redrawn from reference 58.

and also in different parts of the cell, presumably as a result of the variation in availability of these other proteins for interaction with MTs. At one extreme of dynamic instability, activity is high during the early stages of mitosis as MTs grow and shrink rapidly until the end encounters the kinetochore of a chromosome, to which it binds, stabilizing the end. The spindle is fully formed once all chromosomes are connected to the correct number of MTs from each side and coordinated disassembly and movement of the MTs then begin to pull the pairs of chromosomes apart. At the other extreme, long MTs in axons and flagella show very little variation in length. Microtubule associated proteins (MAPs) including MAP1, MAP2 and tau in axons bind to and stabilize MTs and connect them into bundles. In cilia and flagella, over 200 other proteins are involved in stabilizing MTs into a hierarchy of structure and generating the movement that leads to wave-like beating. The axoneme, which forms the core of cilia and flagella, generally contains

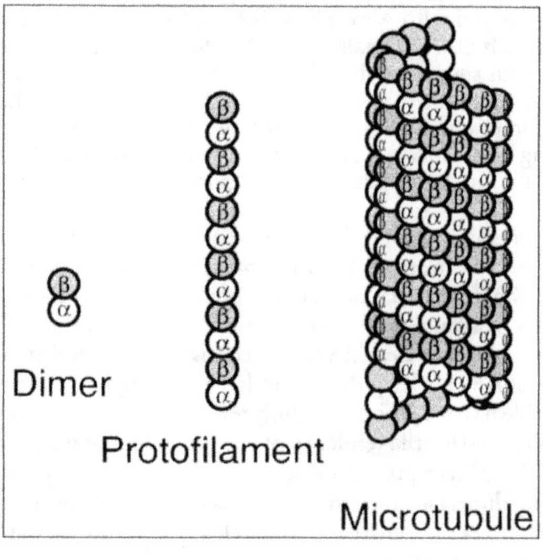

Figure 3. Schematic of microtubule assembly. Alpha-beta tubulin heterodimers form protofilaments which associate in parallel to form the microtubule wall.

two MTs surrounded by nine microtubule doublets, which are structures formed from one intact MT and part of another MT attached to its side. This "9 + 2" arrangement is stabilized by proteins that form radial spokes and other links that bridge neighboring doublets, including sets of dynein motors that produce the movement.

Tubulin Proteins and Their Structure

There are several members of the tubulin family in addition to alpha and beta tubulins. Gamma tubulin[4] is involved with nucleating MTs in the MTOC and probably forms a base ring structure from which the MT grows.[5] Delta and epsilon tubulins appear to be involved in organizing even higher levels of structural complexity and have been localized to the centrosome in the MTOC and the basal body which anchors eukaryotic flagella.[6] Eta tubulin has so far been found just in kinetoplastid protozoans, but there is little doubt that more relatives remain to be discovered.

In almost every organism that contains MTs, tubulin exists in a number of different isoforms expressed from different genes.[7] In humans there are at least eight major isotypes of beta tubulin and seven of alpha. Some yeast and other fungi have only one isotype of beta. The significance of the isotype diversity is still somewhat unclear, although it does seem that some of the isotypes have distinct functions that relate to their presence in specific locations.[8] For example, beta-II -III and -IVa, which show distinctly different dynamics in MTs, are found mainly in neural tissue. Beta-IVb is almost exclusively in flagella, but it is not yet clear whether this is related to the high stability of MTs in flagella or their particular interactions with other proteins.

There has been a remarkable conservation of the amino acid sequence of tubulins throughout the course of evolution. This conservation, which probably is a result of tubulin's interactions with so many other proteins, is particularly remarkable in light of the isotype diversity within species: for example, there is more diversity among human isotypes than between corresponding isotypes of human and other vertebrate tubulins. Figure 4 shows a comparison of sequences of 5 human beta tubulin isotypes compared with one human alpha tubulin and gamma tubulin and with tubulins from other species. Most of the amino acid substitutions that distinguish isotypes are rather conservative, even between species. While there are much larger differences between alpha, beta and gamma tubulins, some regions of the sequence are almost invariant among all tubulins. For example, the "tubulin signature motif" GGGTGSG around residue 140 is present in all tubulins with only very slight substitutions and a long region near the C-terminus is identical in essentially all beta tubulins. The bacterial protein FtsZ was identified from its sequence as a potential homologue of tubulin[9] and indeed its structure is nearly indistinguishable from that of tubulin.[10-12] FtsZ is also involved in cell division, although in a way that is complementary to tubulin's role in segregating chromosomes. FtsZ forms protofilaments which are involved in septation, in a role more like actin's role in eukaryotes.

Figure 5 is a ribbon diagram of the alpha-beta tubulin dimer.[12] Given the close similarity between alpha and beta in primary structure, it is not surprising that the structures of the two monomers are almost identical. Gamma tubulin,[13] as well as FtsZ,[10] also has a structure that is highly similar to alpha and beta tubulins. The structure of tubulin shows the general features of a beta sheet surrounded by alpha helices. The N-terminal half of the sequence forms a classical Rossmann fold that forms the nucleotide binding domain. Within this domain, alpha helices (H1-H6) alternate with beta strands (S1-S6), with each of the loops connecting a strand to the following helix forming part the nucleotide binding pocket. A second domain continues the beta sheet with four strands of mixed polarity (S7-S9) and four more alpha helices (H7-H10). A third domain is formed by a pair of helices (H11, H12) that cross over the first two domains.

The nucleotide-binding pocket is at one end of the dimer, at the interface which is involved in inter-monomer contacts. The non-exchangeability of GTP bound to alpha tubulin is thus clear from the nucleotide being buried at the intra-dimer interface. The exchangeability of the beta-tubulin nucleotide, seen in isolated dimers, is blocked by addition of another dimer to the interface. In addition, a loop from alpha tubulin (T8) fits into the nucleotide pocket of beta, presumably contributing the catalytic function that initiates hydrolysis. The hypothesis that Lys254 of alpha

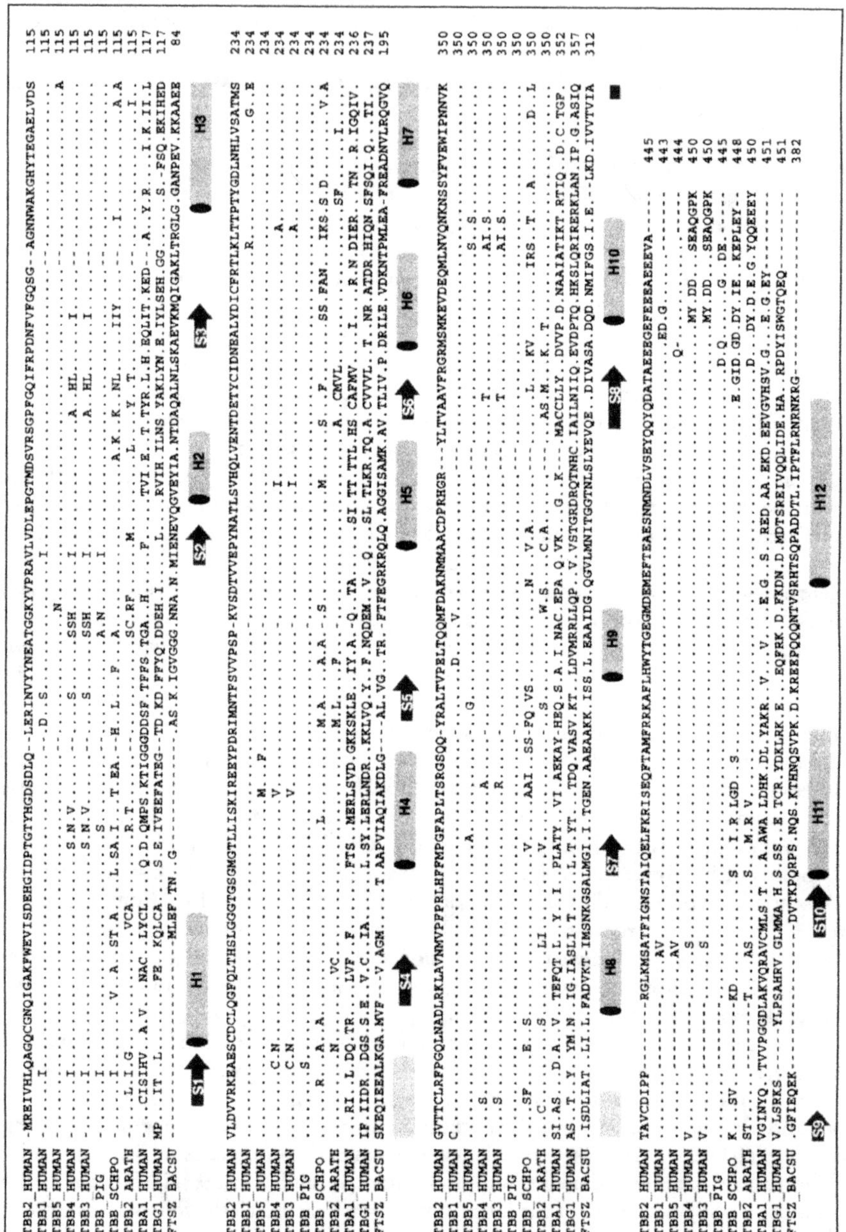

Figure 4. Sequence alignment of 5 human beta tubulin isotypes (TBB1-5) along with beta tubulins from pig (TBB_PIG); the yeast *Schizosaccharomyces pombe* (TBB_SCPO); a plant, *Arabidopsis thaliana* (TBB2_ARATH); human alpha-I (TBA1) and gamma-I (TBG1) tubulins; and FtsZ from *Bacillis subtilis*. Sequences from the Swissprot database were aligned using ClustalX.[59] Dots indicate residues that are the same as in beta-II, dashes are gaps in the sequence. Below the sequences the secondary structures of tubulin are indicated, with cylinders indicating helices and arrows indicating beta strands. The structure can be divided into three domains, with the first domain comprising the first six strands and helices, the second H7-S10 and the third formed by H11 and H12.

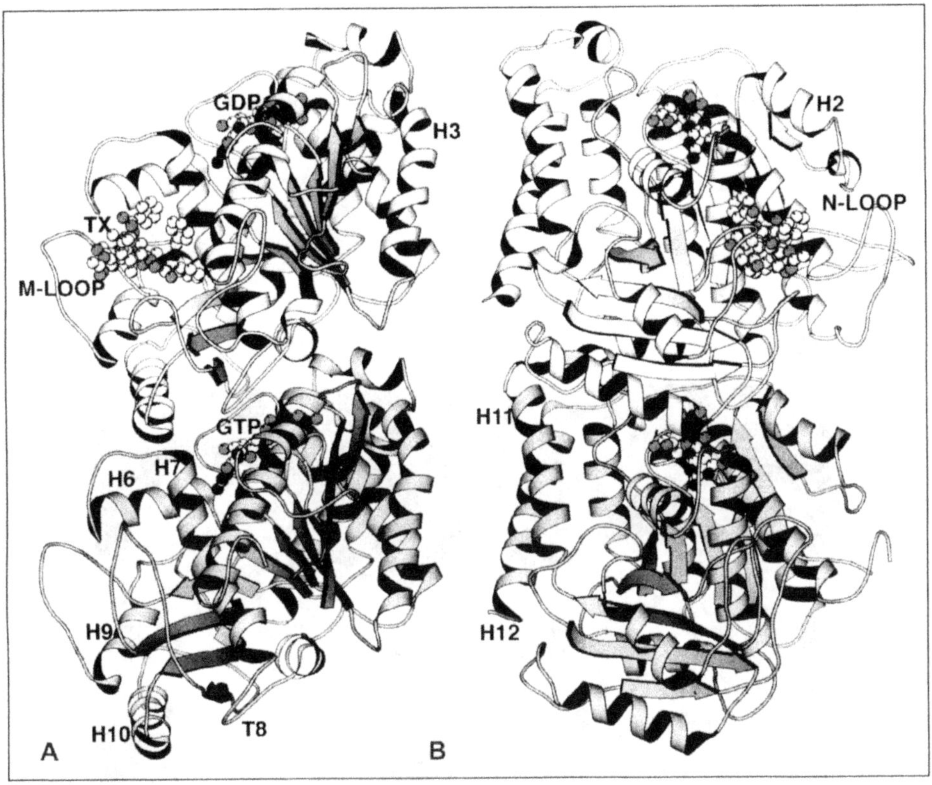

Figure 5. Ribbon diagrams of the tubulin dimer as determined by electron crystallography (PDB structure 1JFF[12]), seen from the inside of the microtubule (A) and rotated 90 degrees (B). The plus end of the microtubule is up. Several of the helices and loops are marked, along with the nucleotides and paclitaxel (TX).

tubulin is the catalytic residue[14] is supported by the observations that mutation of this residue in yeast tubulin produces hyper-stable MTs[15] and that mutation of the corresponding residue in FtsZ abolished hydrolysis.[16]

Electron microscopy (EM) has contributed greatly to the accumulated understanding of tubulin and MTs, starting with the discovery of MTs in the 1960s in thin sections of plastic-embedded cells.[17] With the development of cryo-electron microscopy, several groups have turned to EM for studies of MT structure and interactions with other proteins. The atomic structure of tubulin was first solved by electron crystallography, using two-dimensional (single molecule thick) crystals that form in the presence of zinc ions.[11,12] Much effort has been spent over the years in attempts to produce crystals suitable for X-ray crystallography, but the tendency of pure tubulin to polymerize at high concentration completely thwarted these efforts. In fact tubulin forms a number of different polymers, including sheets, hoops and tubes of varying form, all of which appear to be based on the protofilaments that form MTs. The zinc-induced crystals also contain the protofilaments, but as opposed to the parallel protofilaments of MTs, adjacent protofilaments in the crystals are antiparallel. Thus the crystal structure shows the dimer and its longitudinal interactions within protofilaments, but not the lateral interactions between protofilaments in the MTs.

Cryo-electron microscopy of intact MTs was used to determine how protofilaments interact, as well as to define the orientation of the dimer within the MT. Figure 6A shows one example of the appearance of MTs visualized by cryo-EM. Image averaging and reconstruction techniques can be used to generate a three-dimensional density map from such images. An initial MT model was produced by docking the dimer crystal structure into a three-dimensional map of the MT at a resolution of about 2 nm[18] and later a model with much higher precision was produced with a MT map that was obtained at 0.8 nm resolution[19] and is shown in Figs. 6B,C. Docking clearly identified

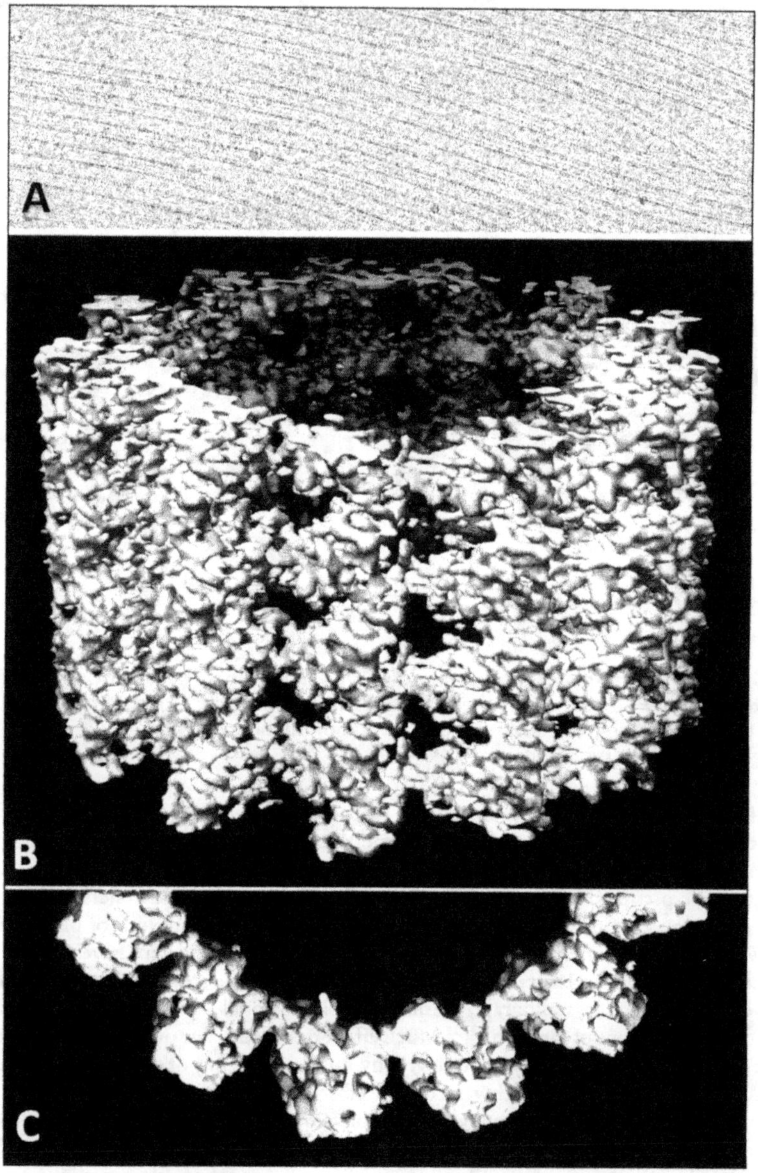

Figure 6. A) Electron micrograph of a preparation of microtubules plunge frozen in liquid ethane to vitrify the water and preserve the native structure with no stain or chemical fixation. B) 3-D reconstruction of microtubule structure at a resolution ~8 Å, obtained from cryo-EM data such as in (a).[19] C) Cross section through the density map in (B), showing the narrow contacts between protofilaments.

the lateral contacts, which consist primarily of a loop—termed the "M-loop"—from one side of the dimer interacting with loops between helix H1 and beta strand S2 (the "N-loop") and with helix H3. The resolution in the most recent maps is still not sufficient to assign individual amino acid interactions and to understand details such as the relative contribution of hydrophobic and charge-charge interactions. However, it is clear how the inter-protofilament interface is flexible enough to adapt to MTs containing different numbers of protofilaments, which can range from nine to16.

While there is no obvious separation of domains within the tubulin monomer, for example as found in actin, there is evidence that the domain structure does have some functional significance. It has been hypothesized that a rotation of the nucleotide domain with respect to the intermediate domain could account for the curvature of the GDP-tubulin protofilaments and rings.[3] X-ray crystal structures of tubulin have finally been obtained by complexing tubulin with stathmin, a protein that is involved in disassembly of MTs, along with drugs that also destabilize MTs.[20,21] According to the current model, stathmin induces the type of curvature found in GDP-protofilaments, pulling tubulin out of the straight form required for polymerization. The crystal structures show a number of distortions with respect to the electron crystallographic structure of the straight conformation. The principal changes are a rotation of the intermediate domain with respect to the nucleotide-binding domain and the presence of kinks of about 12 degrees between each of the monomers, leading to a curved protofilament segment structure. The domain rotations are seen in both alpha and beta monomers and are only slightly smaller in alpha (8 and 11 degrees, respectively). How GTP hydrolysis also induces these types of conformational change is at present unclear, although a number of hypotheses can be suggested.[22] The "core helix", H7, is likely involved in transmitting changes across the molecule. While it contacts the nucleotide at the opposite end from the gamma phosphate, it does represent much of the interface between the nucleotide-binding and intermediate domains and is seen to move by 1.5-2.5 Å in the tubulin-stathmin complexes. It is surprising that crystal structures of FtsZ in several different nucleotide states show none of the domain movements seen in the tubulin-stathmin complex.[23] Thus the force that drives the conformational change still remains to be clarified.

Further information has come from cryo-electron microscopy of tubulin in several other nucleotide states. In the presence of a GTP analog, GMPCPP, it appears that relaxed protofilaments are not straight but slightly curved, with a radius of curvature around 45-60 nm that is achieved by having kinks of about 5 degrees at both inter- and intra-dimer interfaces.[24] In GDP-tubulin rings and spirals, the radius of curvature is typically 15-20 nm, with kinks of 12 and 18 degrees between monomers.[3] While the resolution in these studies is not yet sufficient to resolve details of secondary structure and internal domain rotations, these observations have led to the hypothesis that protofilaments grow from the end of a MT in ribbons or sheets which subsequently close to form the cylindrical MT wall. This model for polymerization is well supported by cryo-EM images that show long extensions of a few protofilaments from the ends of MTs rapidly growing in vitro.[25] However, more recent nanoscale measurements raise questions about whether this model holds under conditions in vivo. Improving the time resolution of these measurements indicates that even during growth phases that appear continuous over times of seconds, such as shown in Figure 2, when a MT is growing up against a restraint such as a cell membrane there is actually a rapid gain and loss of subunits from the MT end that occurs on the time scale of tenths of a second.[26] Simulations support the model that as the pressure against the end of the MT increases the end becomes more blunt, rather than extending a small number of protofilaments as seen in vitro.

In addition to the variety of tubulins that arises from the isotype distribution within a cell, tubulin can undergo several posttranslational modifications.[27] The alpha tubulin sequence ends in a tyrosine and there are specific enzymes that remove and reattach this residue. While the significance of these operations is still unclear, there are antibodies—and presumably other proteins within the cell—that recognize both states. Polygutamylation of residues near the C-terminus of both alpha and beta tubulins is also seen in many cell types. Acetylation of alpha tubulin at Lys40 is a particularly intriguing modification, since this residue is on the luminal surface of the MT. Acetylation is generally associated with stable MTs, although there is still some debate whether acetylation is a cause of stability or a result of the long lifetimes of MTs that are stabilized by other factors.

Small Molecules Affect Microtubule Stability

In addition to the variety of proteins produced by cells to regulate the microtubule cytoskeleton through the cell cycle, a number of organisms produce small molecule compounds that affect MT dynamics, presumably for defensive—or offensive—purposes. Until just a few years ago, these natural compounds were considered to comprise three main classes: colchicine analogues, the vincas and paclitaxel-related compounds. Tubulin was originally identified by its colchicine-binding property.[28] Even earlier, about as soon as MTs were identified, it was recognized that colchicine caused loss of the MT network within cells. Vinblastine was then found to have a similar disruptive effect, although it was soon discovered that vinblastine could also induce a paracrystalline aggregate of tubulin. A large number of compounds related to colchicine and vinblastine have also been found to disrupt MTs, leading to cell cycle arrest and apoptosis. Many attempts followed to develop anticancer drugs that bind to tubulin and there are currently several drugs in use based on vinblastine. The discovery that paclitaxel's cytotoxic effect was due to hyperstabilization of MTs[29] led to another class of anticancer drugs with significantly greater clinical efficacy. The success of paclitaxel as a tubulin-targeted therapy has spawned a great deal of work to develop novel compounds based on the natural products, as well as an extensive search for additional compounds from natural sources. These efforts have not only given oncologists a variety of new options in fighting cancer, but also provided cell biologists and other researchers tools for manipulating the cytoskeleton within cells and MTs in isolation.

Other classes of drugs have been discovered over the years, some of which represent new classes of tubulin-interacting compounds with new binding sites.[30] Docetaxel, a close natural analog of paclitaxel, has many similar properties and the number of synthetic paclitaxel analogues must by now measure in the hundreds.[31] Some of the notable successes have come from designs based on the binding mode of paclitaxel resolved in the crystal structure.[32]

Epothilones, discodermolide and eleutherobin—among others—have been found to bind at or near the paclitaxel site, which is on beta tubulin at the interface between the nucleotide-binding and intermediate domains, adjacent to the M-loop (see Fig. 5). The activity of all of these compounds may arise from inhibiting domain rotation to keep the protofilaments straight, or from stabilizing a particular conformation of the M-loop that favors inter-protofilament interactions. Laulimalide is the first of the MT-stabilizing compounds that appears to bind at a site different from paclitaxel's.[33] Benzimidazoles,[34] including benomyl and thiabendazole, have been used both in research to disrupt MTs and in agriculture as antifungals and antihelminths. It was evident from competitive binding and other studies that the benzimidazoles bind to beta tubulin but at a site different from the other compounds and this site has been fairly well localized through a variety of mutational studies.[35] Dinitroanilines, particularly oryzalin and trifluralin, have been widely used as herbicides and it was eventually found that they act by disrupting MTs.[36] Dinitroanilines are the first of the MT-affecting compounds identified as binding to alpha tubulin.[37] Figure 7 shows schematically the relationships among the various drug binding sites.

All of these compounds show the ability to either induce or inhibit tubulin polymerization. It has become apparent, though, that they produce cytotoxicity at concentrations significantly below that required to alter the mass of MTs in the cell. Dynamic instability can be substantially affected at these lower concentrations[38] and the general paradigm has evolved that cells are driven into apoptosis when the dynamic processes associated with searching for kinetochores are altered and checkpoint mechanisms recognize that chromosome capture has not been successful.

Microtubule Motors

The most well studied role for microtubules is as tracks for motor molecules that carry cargo back and forth within the cell. Kinesin motors, of which there are at least 14 families so far identified, generally move toward the MT plus end.[39] Dyneins and a few proteins in the kinesin family

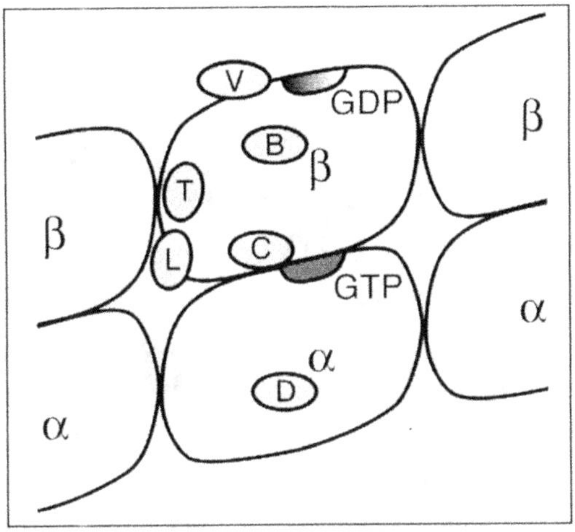

Figure 7. Schematic representation of locations of the various drug binding sites on tubulin, with the dimer viewed in the same orientation as in Fig. 5A. Classes and representative members include: T) paclitaxel and most other MT-stabilizing compounds, including docetaxel, discodermolide, epothilone. Location determined in electron crystallographic structure.[12] L) laulimalide, peloruside. This location is somewhat speculative, inferred from computational modeling studies (J. Nettles, unpublished). C) colchicine and related compounds including podophyllotoxin, combretastatin. Location determined in X-ray crystal structure.[20] V) vinblastine and related compounds including vinorelbine, maytansine, phomopsin A. Location determined in X-ray crystal structure.[21] B) benzimidazoles, e.g., benomyl, thiabendazole, carbendazim. Location inferred from effects of mutations that cluster around the site and are associated with resistance to these compounds.[35] D) dinitroanilines , including trifluralin, oryzalin. Location inferred from mutations associated with resistance.[37]

such as Kar3 move toward the minus end. All of the motors studied to date use ATP hydrolysis to power their movements along the MT. Electron microscopy has shown that both kinesin and dynein bind to approximately the same region on the MT, along the ridge of the protofilaments.[40] X-ray crystallography has been used to determine structures of a number of different kinesins in several nucleotide states,[41-43] but not in association with MTs. Recently cryo-EM studies of kinesin-decorated MTs have reached a resolution sufficient to show enough detail in secondary structure to allow very precise docking of X-ray crystal structures of kinesins into 3-D density maps and even to show some structural differences between MT-bound kinesin and the unbound motors that were studied in the crystals.[44-46] Figure 8 is a 3-D density map derived by cryo-EM of kinesin-decorated MTs, into which crystal structures of tubulin and kinesin have been docked. At the resolution of this map, around 9 Å, the docking is precisely constrained by the map. Work of this type has produced snapshots of the kinesin-MT complex in several nucleotide states and is giving a good understanding of how the ATP hydrolysis cycle relates to MT binding and how the hydrolysis-induced conformational changes in kinesin lead to movement along the MT. Similar work with dynein has been slower because of the much greater size of dynein, as well as its apparently flexible attachment to the MT.

The same approach with other proteins that bind statically but tightly to MTs should eventually provide insights on the general modes of the association. For example, results with MAP2 and tau already suggest a general pattern of binding along the protofilament crest in the same region as the motors[47] and also suggests that proteins may be able to bind across the MT wall to reach sites on the luminal surface as well as the exterior.[48]

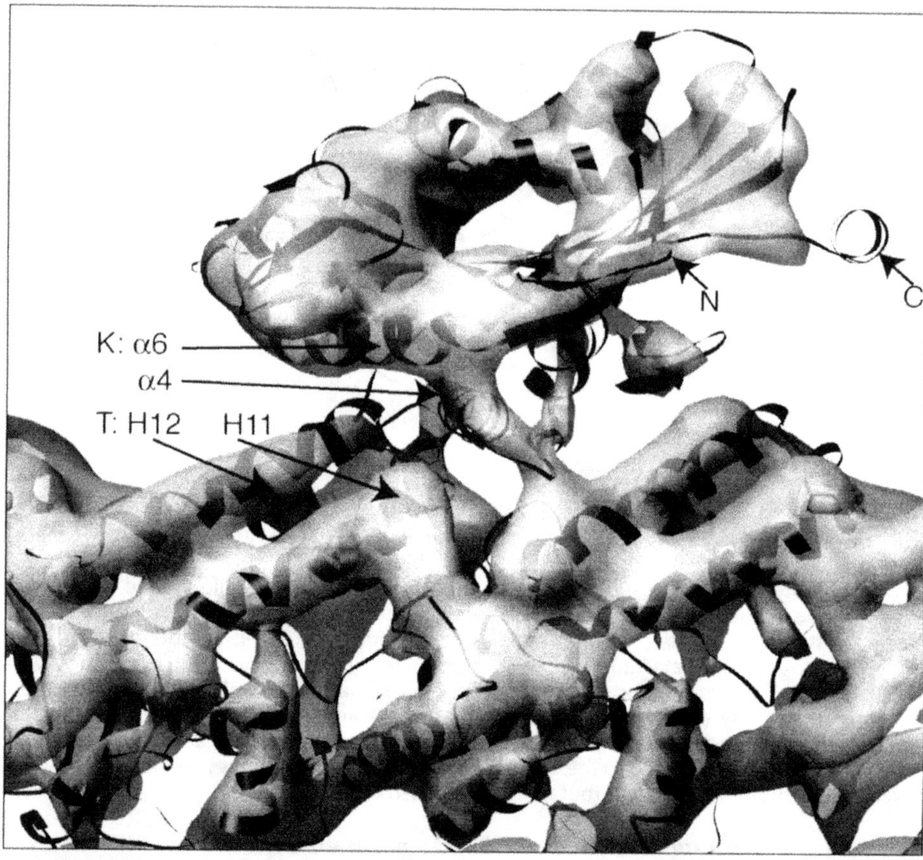

Figure 8. Three-dimensional map of kinesin-decorated microtubule determined by cryo-electron microscopy, with crystal structures of kinesin and tubulin docked into the density.[46] The resolution of the density map is about 9 Å, and the clear definition of helix densities allows precise positioning of the crystal structures. Alpha helices 4 and 6 in kinesin are marked (α4, α6) as well as helices 11 and 12 in tubulin (H11, H12). The N- and C-termini of kinesin are also marked. The C-terminal helix shows no density, presumably because its position is highly variable. The plus end of the microtubule is to the right.

Mechanical Properties of Microtubules

The other principal function of MTs has to do with maintaining cell shape, including some involvement in cell movement. The MT network works in concert with actin and other components of the cytoskeleton in this function and the general model is that MTs provide an expansive force which is balanced by tension from the actin network.[49] There are proteins that connect MT ends to cell membranes, but most of this expansion force derives from microtubules simply pushing towards the cell periphery as they grow. The amount of force delivered to the membrane depends on many factors such as the angle of contact and specific interaction sites, but a general impression of the force available comes from the observation that a force of 2.8 pN on the end of a MT reduces the rate of growth by only a factor of two.[26] To further understand the forces that MTs could provide, many efforts have been made to measure mechanical properties of MTs.[50,51] There are differences in these measurements, but in general they suggest that the flexural rigidity of a MT is comparable to that of Lucite, around 2-5 × 10^{-24} Nm^2. Measurements by atomic force microscopy indicate that the strength of the wall itself is reasonably well modeled by assuming it to be made of a uniform material with a Young's modulus of about 0.2 GPa, compatible with the flexural rigidity.[52] With its hollow structure, the MT is engineered to maximize stiffness for a given amount of material and MTs are certainly the stiffest components of the cytoskeleton. Their mechanical support within the cell is further enhanced by interactions with other cytoskeletal components that crosslink or bundle them.

Nanotechnology Applications

The facts that MTs are such stiff structures with a size on the molecular scale and that they can both be easily manipulated and interact predictably with motor molecules has inspired a great deal of thinking about potential applications on the nano scale. While most of these inspirations are still in the proof-of-concept stage, it is already clear that tubulin may have as rich a future on nanotechnology as it has had in clinical medicine.

Some of these ideas take advantage of tubulin's inherent dynamic instability. As early as a decade ago it was shown that asters of MTs grown from isolated MTOCs in microfabricated chambers would be pushed in both predictable and stochastic ways as MTs grow and shrink.[53] More recently, simulations have been used to demonstrate how this behavior could be used to sort and deliver cargo on the micrometer scale.[54]

Much of the work in the nanotechnology context is done with paclitaxel-stabilized MTs, which have a useful lifetime of at least several days under many conditions. (This is in marked contrast to unpolymerized tubulin, which has a half life on the order of hours.) In some interesting work that relates to electrophoretic manipulation of MTs, it was shown that the mobility of MTs does not follow Stokes hydrodynamics, but that MTs have a higher anisotropy so that they move more easily than expected along the axial direction. In addition, this axial motion is independent of length.[55] From this work a surface charge of about 23 electrons per dimer was calculated, complementing previous estimations of high electric and magnetic dipole moments for MTs. Thus MTs should be amenable to manipulation by both electric and magnetic fields.

Biophysicists studying kinesin movement along MTs have developed several assays, one of which uses high-sensitivity fluorescence imaging to follow single labeled kinesin molecules along MTs and another of which (the "MT gliding assay") follows MTs as they move along kinesin molecules that are immobilized on a substrate. These approaches have been used in several nanotechnology applications, both to demonstrate controlled transport of fluorescent cargoes attached to kinesin motors[56] and also to investigate control of the MT movement itself. In an experiment where MTs were attached to the floor of microfabricated channels, it was found that particles attached to kinesin could move at around 2 μm/sec, comparable to the velocity of kinesins in cells and that individual particles could be moved in defined channels over distances of millimeters.[57] Possibilities abound for using this approach to sort individual molecules, to deliver samples to sensors and in a myriad of other applications limited only by vision of the investigator.

Conclusion

Microtubules perform many functions in cells relating to establishing cell shape, transport of material within the cell and movement of cells. The fact that these functions are essential to the life of the cell and that small ligands can disrupt normal activity has made tubulin a successful target in anti-cancer therapy, as well as in other aspects of clinical medicine and agriculture. The physical properties of MTs that allow them to perform these functions, along with the interactions of MTs with many other proteins including motor molecules, has inspired applications of MTs as structural and functional elements in nanotechnology. Thus in the half century since their discovery, MTs have developed from being interesting and essential subcellular structures with fascinating physical properties to the core of several medical and technological applications. It will be interesting to see how this development continues, both in vivo and in vitro.

References

1. Mitchison T, Kirschner M. Dynamic instability of microtubule growth. Nature 1984; 312:237-242.
2. Mandelkow E-M, Mandelkow E, Milligan RA. Microtubule dynamics and microtubule caps: a time-resolved cryo-electron microscopy study. J Cell Biol 1991; 114:977-991.
3. Wang HW, Nogales E. Nucleotide-dependent bending flexibility of tubulin regulates microtubule assembly. Nature 2005; 435:911-5.
4. Oakley BR, Oakley CE, Yoon Y et al. Gamma-tubulin is a component of the spindle pole body that is essential for microtubule function in Aspergillus nidulans. Cell 1990; 61:1289-1301.
5. Schiebel E. gamma-tubulin complexes: binding to the centrosome, regulation and microtubule nucleation. Curr Opin Cell Biol 2000; 12:113-8.
6. Dutcher SK. Long-lost relatives reappear: identification of new members of the tubulin superfamily. Curr Opin Microbiol 2003; 6:634-40.
7. Cleveland DW, Sullivan KF. Molecular biology and genetics of tubulin. Annu Rev Biochem 1985; 54:331-365.
8. Carpenter EJ, Huzil JT, Luduena RF et al. Homology modeling of tubulin: influence predictions for microtubule's biophysical properties. Eur Biophys J 2006; 36:35-43.
9. RayChaudhuri D, Park JT. Escherichia coli cell-division gene FtsZ encodes a novel GTP-binding protein. Nature 1992; 359:251-254.
10. Löwe J, Amos LA. Crystal structure of the bacterial cell division protein FtsZ. Nature 1998; 391:203-206.
11. Nogales E, Wolf SG, Downing KH. Structure of the αβ tubulin dimer by electron crystallography. Nature 1998; 391:199-203.
12. Löwe J, Li H, Downing KH et al. Refined structure of alpha beta-tubulin at 3.5 Å resolution. J Mol Biol 2001; 313:1045-1057.
13. Aldaz H, Rice LM, Stearns T et al. Insights into microtubule nucleation from the crystal structure of human gamma-tubulin. Nature 2005; 435:523-7.
14. Nogales E, Downing K, Amos L et al. Tubulin and FtsZ form a distinct family of GTPases. Nature Struct Biol 1998; 5:451-458.
15. Anders KR, Botstein D. Dominant-lethal alpha-tubulin mutants defective in microtubule depolymerization in yeast. Mol Biol Cell 2001; 12:3973-3986.
16. Dai K, Mukherjee A, Xu Y et al. Mutations in FtsZ that confer resistance to SulA affect the interaction of FtsZ with GTP. J Bacteriol 1994; 175:130-136.
17. Ledbetter MC, Porter KR. A "microtubule" in plant fine structure. J Cell Biol 1963; 19:239-250.
18. Nogales E, Whittaker M, Milligan RA et al. High resolution model of the microtubule. Cell 1999; 96:79-88.
19. Li H, DeRosier D, Nicholson W et al. Microtubule structure at 8 Å resolution. Structure 2002; 10:1317-1328.
20. Ravelli RBG, Gigant B, Curmi PA et al. Insight into tubulin regulation from a complex with colchicine and a stathmin-like domain. Nature 2004; 428:198-202.
21. Gigant B, Wang C, Ravelli RB et al. Structural basis for the regulation of tubulin by vinblastine. Nature 2005; 435:519-22.
22. Amos L, Löwe J. How Taxol stabilizes microtubule structure. Chem Biol 1999; 6:R65-R69.
23. Oliva MA, Trambaiolo D, Lowe J. Structural insights into the conformational variability of FtsZ. J Mol Biol 2007; 373:1229-42.
24. Nogales E, Wang HW. Structural mechanisms underlying nucleotide-dependent self-assembly of tubulin and its relatives. Curr Opin Struct Biol 2006; 16:221-9.
25. Chrétien D, Fuller SD, Karsenti E. Structure of growing microtubule ends: two-dimensional sheets close into tubes at variable rates. J Cell Biol 1995; 129:1311-1328.
26. Schek HT 3rd, Gardner MK, Cheng J et al. Microtubule assembly dynamics at the nanoscale. Curr Biol 2007; 17:1445-55.
27. Ludueña RF. The multiple forms of tubulin: different gene products and covalent modifications. Int Rev Cyt 1998; 178:207-275.
28. Weisenberg RC, Borisy GG, Taylor EW. The colchicine-binding protein of mammalian brain and its relation to microtubules. Biochemistry 1968; 7:4466-4479.
29. Horwitz SB. Mechanism of action of taxol. TiPS 1992; 13:134-136.
30. Hamel E, Lin CM, Flynn E et al. Interactions of 2-methoxyestradiol an endogenous mammalian metabolite, with unpolymerized tubulin and with tubulin polymers. Biochemistry 1996; 35:1304-1310.
31. Kingston DG, Newman DJ. Taxoids: cancer-fighting compounds from nature. Curr Opin Drug Discov Devel 2007; 10:130-44.
32. Ganesh T, Yang C, Norris A et al. Evaluation of the tubulin-bound paclitaxel conformation: synthesis, biology and SAR studies of C-4 to C-3' bridged paclitaxel analogues. J Med Chem 2007; 50:713-25.

33. Mooberry SL, Tien G, Hernandez AH et al. Laulimalide and isolaulimalide, new paclitaxel-like microtubule stabilizing agents. Cancer Research 1999; 59:653-660.
34. Gupta RS. Cross-resistance of nocodazole-resistant mutants of CHO cells toward other microtubule inhibitors: similar mode of action of benzimidazole carbamate derivatives and NSC 181928 and TN-16. Mol Pharmacol 1986; 30:142-8.
35. Jung MK, Wilder IB, Oakley BR. Amino acid alterations in the benA (beta-tubulin) gene of Aspergillus nidulans that confer benomyl resistance. Cell Motil Cytoskel 1992; 22:170-174.
36. Anthony RG, Hussey PJ. Double mutation in Eleusine indica alpha-tubulin increases the resistance of transgenic maize calli to dinitroaniline and phosphorothioamidate herbicides. Plant J 1999; 18:669-674.
37. Ma C, Li C, Ganesan L et al. Mutations in {alpha}-tubulin confer dinitroaniline resistance at a cost to microtubule function. Mol Biol Cell 2007; 18: 4711-20
38. Wilson L, Panda D, Jordan MA. Modulation of microtubule dynamics by drugs: A paradigm for the actions of cellular regulators. Cell Struct Funct 1999; 24:329-335.
39. Miki H, Okada Y, Hirokawa N. Analysis of the kinesin superfamily: insights into structure and function. Trends Cell Biol 2005; 15:467-76.
40. Mizuno N, Toba S, Edamatsu M et al. Dynein and kinesin share an overlapping microtubule-binding site. EMBO J 2004; 23:2459-67.
41. Kull FJ, Sablin EP, Lau R et al. Crystal structure of the kinesin motor domain reveals a structural similarity to myosin. Nature 1996; 380:550-554.
42. Sack S, Muller J, Marx A et al. X-ray structure of motor and neck domains from rat brain kinesin. Biochemistry 1997; 36:16155-65.
43. Kikkawa M, Sablin EP, Okada Y et al. Switch-based mechanism of kinesin motors. Nature 2001; 411:439-45.
44. Kikkawa M, Hirokawa N. High-resolution cryo-EM maps show the nucleotide binding pocket of KIF1A in open and closed conformations. EMBO J 2006; 25:4187-94.
45. Hirose K, Akimaru E, Akiba T et al. Large conformational changes in a kinesin motor catalyzed by interaction with microtubules. Mol Cell 2006; 23:913-23.
46. Sindelar CV, Downing KH. The beginning of kinesin's force-generating cycle visualized at 9-A resolution. J Cell Biol 2007; 177:377-85.
47. Al-Bassam J, Ozer RS, Safer D et al. MAP2 and tau bind longitudinally along the outer ridges of microtubule protofilaments. J Cell Biol 2002; 157:1187-1196.
48. Kar S, Fan J, Smith MJ et al. Repeat motifs of tau bind to the insides of microtubules in the absence of taxol. EMBO J 2003; 22:70-77.
49. Brangwynne CP, MacKintosh FC, Kumar S et al. Microtubules can bear enhanced compressive loads in living cells because of lateral reinforcement. J Cell Biol 2006; 173:733-41.
50. Gittes F, Mickey B, Nettleton J et al. Flexural rigidity of microtubules and actin filaments measured from thermal fluctuations in shape. J Cell Biol 1993; 120:923-34.
51. Felgner H, Frank R, Schliwa M. Flexural rigidity of microtubules measured with the use of optical tweezers. J Cell Sci 1996; 109(Pt 2):509-16.
52. Schaap IA, Carrasco C, de Pablo PJ et al. Elastic response, buckling and instability of microtubules under radial indentation. Biophys J 2006; 91:1521-31.
53. Holy TE, Dogterom M, Yurke B et al. Assembly and positioning of microtubule asters in microfabricated chambers. Proc Natl Acad Sci USA 1997; 94:6228-31.
54. Bouchard AM, Warrender CE, Osbourn GC. Harnessing microtubule dynamic instability for nanostructure assembly. Phys Rev E Stat Nonlin Soft Matter Phys 2006; 74:041902.
55. van den Heuvel MG, de Graaff MP, Lemay SG et al. Electrophoresis of individual microtubules in microchannels. Proc Natl Acad Sci USA 2007; 104:7770-5.
56. Muthukrishnan G, Hutchins BM, Williams ME et al. Transport of semiconductor nanocrystals by kinesin molecular motors. Small 2006; 2:626-30.
57. Cheng LJ, Kao MT, Meyhofer E et al. Highly efficient guiding of microtubule transport with imprinted CYTOP nanotracks. Small 2005; 1:409-414.
58. Detrich HW, Parker SK, Williams RC et al. Cold adaptation of microtubule assembly and dynamics—Structural interpretation of primary sequence changes present in the alpha- and beta-tubulins of Antarctic fishes. J Biol Chem 2000; 275:37038-37047.
59. Chenna R, Sugawara H, Koike T et al. Multiple sequence alignment with the Clustal series of programs. Nucleic Acids Res 2003; 31:3497-500.

Chapter 8

Silks Produced by Insect Labial Glands

František Sehnal* and Tara Sutherland

Abstract

Insect silks are secreted from diverse gland types; this chapter deals with the silks produced by labial glands of Holometabola (insects with pupa in their life cycle). Labial silk glands are composed of a few tens or hundreds of large polyploid cells that secrete polymerizing proteins which are stored in the gland lumen as a semi-liquid gel. Polymerization is based on weak molecular interactions between repetitive amino acid motifs present in one or more silk proteins; cross-linking by disulfide bonds may be important in the silks spun under water. The mechanism of long-term storage of the silk dope inside the glands and its conversion into the silk fiber during spinning is not fully understood. The conversion occurs within seconds at ambient temperature and pressure, under minimal drawing force and in some cases under water. The silk filament is largely built of proteins called fibroins and in Lepidoptera and Trichoptera coated by glue-type proteins known as sericins. Silks often contain small amounts of additional proteins of poorly known function. The silk components controlling dope storage and filament formation seem to be conserved at the level of orders, while the nature of polymerizing motifs in the fibroins, which determine the physical properties of silk, differ at the level of family and even genus. Most silks are based on fibroin β-sheets interrupted with other structures such as α-helices but the silk proteins of certain sawflies have predominantly a collagen-like or polyglycine II arrangement and the silks of social Hymenoptera are formed from proteins in a coiled coil arrangement.

Introduction

Silks are (1) ectodermal secretions that (2) are stored as a hydrated jelly within cells or, more often, in multicellular cavities; (3) the gels polymerise into water-insoluble filaments during passage to the external environment. Materials that meet this definition occur in all terrestrial arthropod subphyla: Chelicerata (in three orders of Arachnida, i.e., spidermites, pseudoscorpions and spiders), Myriapoda (in Symphyla, Pauropoda and primitive millipedes, absent in the centipedes) and Hexapoda (Fig. 1). Silk is always secreted by specialised ectodermal cells that make up a silk gland.[1] Rudall and Kenchington distinguished eight structural silk categories based on X-ray diffraction data and suggested that all silk proteins are phylogenetically homologous.[2] This view is no longer tenable. The diversity of silk-producing taxa, their silk glands and the silk composition suggest that silk has evolved independently a number of times as a product of the cuticle-secreting cells. The cuticle is composed of chitin and a blend of proteins, many of which contain repeats of short amino acid motifs;[3] several of such proteins could be the precursors of silks.

Insect silks are produced by a variety of dermal glands, Malpighian tubuli, colleterial sex glands, the gut and the labial glands (Table 1). The latter typically produce saliva but half of each gland in the Psocoptera and the entire glands in the larvae of many Holometabola, are specialized for the production of silk. Labial silk glands develop during embryogenesis as ectodermal tubular invaginations composed of several tens to a few hundred cells. They grow by polyploidization, extend into the abdomen and often reach a large size, second only to the gut and the fat body.

*Corresponding Author: František Sehnal—Biology Centre ASCR, Branišovská 31, 370 05 České Budějovice, Czech Republic. Email: sehnal@bc.cas.cz

Fibrous Proteins, edited by Thomas Scheibel. ©2008 Landes Bioscience.

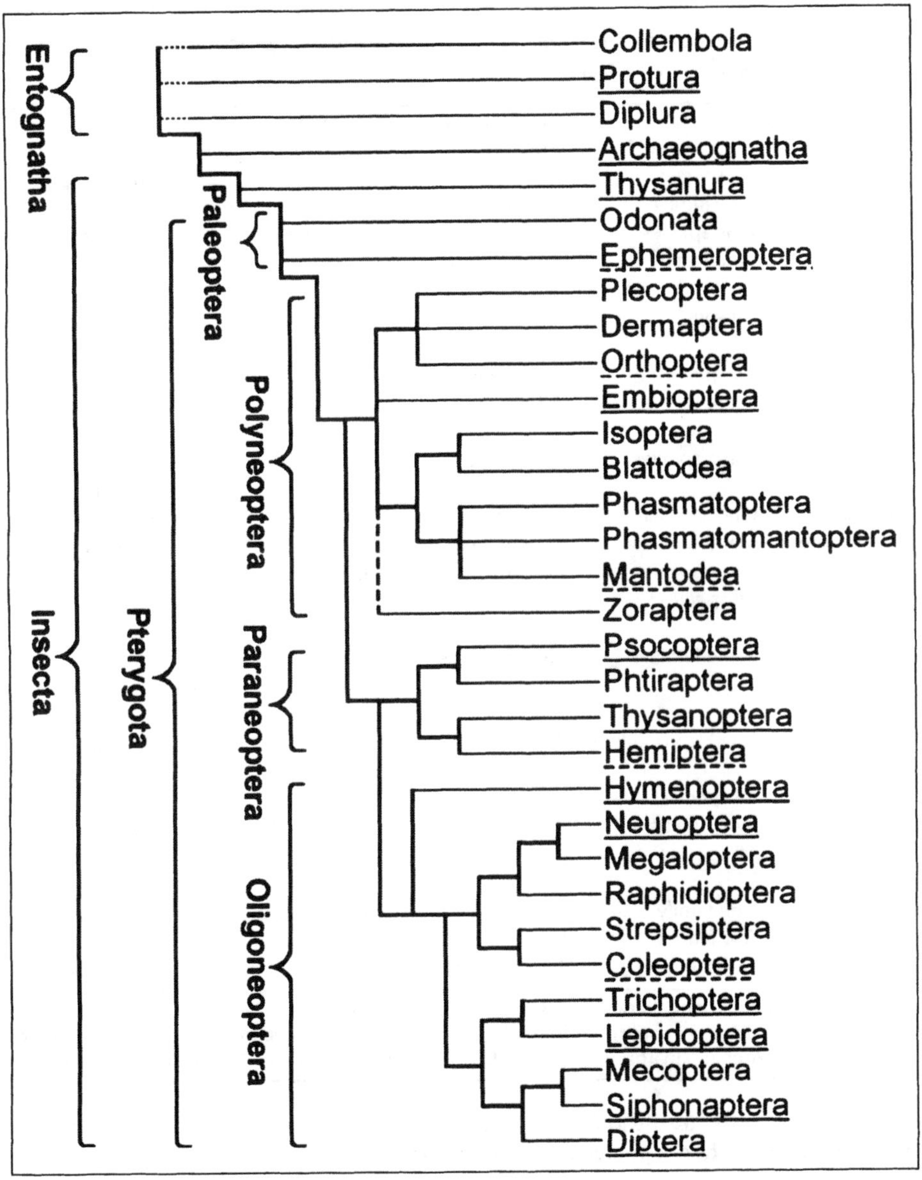

Figure 1. Approximate evolutionary relationships and the distribution of silk production in the orders of Hexapoda. Silk production is common in orders underlined with a solid line and exceptional (limited to a small number of species) in those underlined with a dashed line.

Most information on insect silks concerns the secretions of labial glands that are the subject of this review; little is known on the other silk types.[1,4] Holometabola are monophyletic (Fig. 1) but the over-all composition of the silky secretions from their labial glands has greatly diversified (Table 2). The mechanism of silk secretion, silk composition and the silk structure has been examined most systematically in Lepidoptera, being stimulated by the interests of commercial sericulture. Since

Table 1. Silk production in different Hexapod orders (developmental stage, type of glands and silk function)

Protura	??	Adults (Larvae?)	Abdominal glands	??
Archaeognatha	All species?	Adults	Dermal glands (cerci)	Sperm support
Thysanura	All species?	Adults	Dermal glands (cerci)	Sperm support
Ephemeroptera	One family	Larvae	Malpighian tubuli	??
Odonata	One species	Adults	Unknown	Egg coverings
Orthoptera[69]	One Family	Adults and larvae	Maxillary glands	Nests and burrows
Embioptera	All species	Larvae and adults	Dermal glands (tarsi)	Tunnels and egg coverings
Mantodea	All species	Adult females	Colleterial glands	Ootheca
Psocoptera	Some species	Adults	Labial glands	Egg covering and retreats
Thysanoptera[70]	Some species	Adults and Larvae	Anal glands in adults	Domiciles or cocoons
Hemiptera[71]	One species	Unknown	Unknown	Shelter
Hymenoptera	Many species	Grown larvae	Labial glands	Cocoon
Neuroptera	Some species	Grown larvae	Malpighian tubuli	Cocoon
	All species	Adult females	Colleterial glands	Egg stalk or cover
Coleoptera	Some families	Grown larvae	Malpighian tubuli or Peritrophic membrane	Cocoon, Underwater tunnels
	Few Hydrophilidae	Adult females	Colleterial glands	Floating rafts for eggs
Trichoptera	All species	Larvae	Labial glands	Diverse
Lepidoptera	Most species	Larvae	Labial glands	Diverse
Siphonaptera	All species?	Grown larvae	Labial glands	Nests? and Cocoon
Diptera	One family	Larvae	Labial glands	Prey capture
	Few families	Larvae	Labial glands	Tunnels
	One family	Tarsal glands	Dermal glands (tarsi)	Prey package

Data were drawn from Sehnal and Akai[1], Craig[4] and other publications cited either in the table or in the text (for orders printed in bold). Additional information was obtained from our expert colleagues J. Rusek (Protura, Archaeognatha, Thysanura), J. Trueman (Odonata) and P. Švácha (Coleoptera).

lepidopteran silk is often used as a prototype of the other silks, it is appropriate to begin this chapter with a description of the silk production in caterpillars.

The Silks of Caterpillars (Larvae of Lepidoptera)

Spinning from the labial glands of larvae probably evolved in the ancestor of the Lepidoptera and the Trichoptera, more than 250 million years ago. Caterpillars of some species spin virtually continuously and live in silky tubes or domiciles; other species produce just a small pad or a girdle that provides support during moulting. Many species spin durable cocoons in which the larvae pupate. Silk composition has been analysed in some detail in the suprafamilies Yponomeutoidea, Pyraloidea and Bombycoidea. The conclusions are probably valid for the entire clade Ditrysia that harbours 98% of about 160,000 described lepidopteran species.[5]

The silk fiber of Ditrysia is a highly organised structure composed of several proteins that are derived from the posterior and middle silk gland sections (PSG and MSG, respectively). The fiber

Table 2. Dominant amino acids in the major silk proteins produced from the labial glands of Holometabolous insects

	Hymenoptera			Diptera-Nematocera		Trichoptera		Lepidoptera			
	Cg	**Vs**	**Am**	**Cte BR1**	**Cth 220**	**Ha**	**Ld**	**Ye**	**Gm**	**Bm**	**Ap**
Gly						21.7%	24.6%	27.7%	31.6%	45.9%	27.3%
Ala		34.6%	32.7%					26.3%	23.8%	30.2%	43.1%
Ser	32.6%	22.7%		21.4%		17.8%	17.3%	17.8%	18.1%		
Lle, Leu, Val			15.1%			20.5%	21.5%		16.0%		
Arg, His, Lys				39.0%			16.7%				
Asn	38.4%										
Cys					17.5%						

Analysis is based on 254 residues from *Cotesia glomerata* (Cg),[61] 1652 (four full length proteins) from *Vespa simillima* (Vs)[67] and 1243 (four full length proteins) from *Apis mellifera* (Ap)[66] silk proteins. Silk proteins BR1 (210 residues)[72] and sp220 (full sequence, 1704 residues)[73] were analysed in *Chironomus tentans* (Cte) and *Ch. thummi* (Cth), respectively. The composition of H-fibroins is presented in the caddisflies *Hydropsyche angustipennis* (Ha, 1042 residues) and *Limnephilus decipiens* (Ld, 516 residues),[44] and in the moths *Yponomeuta evonymella* (Ye, 955 residues),[26] *Galleria mellonella* (Gm, 1277 residues),[38] *Bombyx mori* (full sequence of 5263 residues),[35] and *Antheraea pernyi* (Ap, full sequence of 2639 residues).[28]

consists of two filaments (one from each gland) that are polymers of a large protein called heavy chain fibroin (H-fibroin, 350-500 kDa in different species). It is produced in the PSG along with two other proteins, light chain fibroin (L-fibroin, cca 25 kDa) and P25 glycoprotein (also called fibrohexamerin) that occurs in two forms due to various levels of glycosylation (27-31 kDa). These three proteins are assembled into elementary units in the endoplasmic reticulum,[6] form microfibers in the Golgi vacuoles and are released as secretory granules into the gland lumen,[7] where they accumulate as a highly concentrated gel. The gel moves into the MSG and becomes enveloped by several layers of sericins. Major sericins of the domestic silkworm, *Bombyx mori*, include a 150 kDa protein produced in the most distal part of MSG, a 400 kDa protein from the central and a 250 kDa protein from the proximal MSG region.[8] Silk proteins polymerise into a fiber during spinning when they pass through the anterior silk gland section and the spinneret. Filament polymerisation is based on cross-linking of aligned and closely packed H-fibroin molecules. This process apparently requires the withdrawal of water and application of a shearing force. Sericins polymerise with a delay: inner sericins (from the distal MSG) seal the filament doublet into a single fiber and outer sericins glue the fibers to substrates or to one another during cocoon construction.

Commercial silks are products of the domestic silkworm *Bombyx mori*, several silkmoth species (most of them from the genus *Antheraea*) and a few other moths whose larvae spin large and closed cocoons. To release the fiber, cocoons are soaked in hot and slightly alkaline water that dissolves the outer sericin layer. Several fibers with a sticky surface are reeled together into a raw silk thread. Measurements of the physical properties and structural silk analyses by X-ray diffraction and other techniques are typically done on fibers liberated from the cocoons in hot water bath; it is assumed that the established values characterise the filaments, i.e., primarily the H-chain fibroin. Sericins and additional,[9,10] minor silk components are not considered.

At least 6 and as many as 15 sericin-type proteins were extracted from the silk of *B. mori*.[11,12] Some of them may be based on identical peptides but differ in the degree of glycosylation. *B. mori* may harbour up to five sericin genes[13] but only *ser1* and *ser2* have been identified. The *ser1* gene of 9 exons is expressed in the distal and central MSG and yields 4-5 mRNAs by differential splicing.[14,15] A large central exon encodes 60 copies of a 38 amino acid repeat.[16] Recombinant proteins based on a common version of this repeat (SRTSGGTSTYGYSSSHRGGSVSSTGSSSNTDSSTKNAG) were showed to self-assemble into β-sheets and crystallites.[17] Based on this observation, filament sealing into a fiber may be attributed to hydrogen bonding between serine residues of the H-fibroin and serine residues of the inner-layer sericins. Recent unpublished analysis of *ser2*, which is spliced into two mRNAs,[18,19] disclosed 11 exons, two of which encode multiple copies of the repeat ALTEKAKPNDRSPSHKDT. The secondary structure and the manner of presumed Ser2 polymerisation are unknown. It is not excluded that the Ser2 proteins actually hinder polymerisation and hold the cocoon fibers together by electrostatic forces that are destroyed by cooking in the alkaline environment. The *ser3* gene of 3 exons incodes repeats composed of an 86- and an 8-amino acids motif, respectively.[75] Genes similar to *ser1* and *ser2* in the overall structure and the splicing patterns, but different in repetitive sequences, were found in the wax moth *Galleria mellonella* of Pyraloidea.[20] Sericins have not been sufficiently examined in any other species.

The investigations on *B. mori* showed that H-fibroin, L-fibroin and P25 assemble in the endoplasmic reticulum in a ratio of 6:6:1 into elementary secretory units.[21] H-fibroin interacts with the two other proteins by its nonrepetitive N- and C-termini. These terminal sequences are conserved across Lepidoptera and Trichoptera (Fig. 2). The N-terminus harbours an amphiphilic proline-flanked region of 35 residues and the C-terminal 26-27 residues include three Cys (only one in the Saturniidae family) in conserved positions. The redox state of the endoplasmic reticulum[22] favours formation of disulfide bonds: four internal in P25, one within L-fibroin, one within H-fibroin and one linking Cys_{170} of the L-fibroin to Cys_{-22} in the H-fibroin C-terminus.[23] Disulfide linkage between L-fibroin and H-fibroin is indispensable for the secretion of both proteins.[24] Non-covalent interaction of P25 with the H-fibroin N-terminus facilitates transport and secretion of the highly insoluble H-fibroin/L-fibroin heterodimers.[25] At the same time, L-fibroin plays a protective role in P25 glycosylation—full deglycosylation causes disintegration of the elementary

A

```
          v         v         v         v         v         v         v
Ro   MRAIFLLILVCSLQINIAGSSSVIDKLEDLLTHGHHGHHWEGTDGLHEKLLQEDDVIEANSKGEIIEKIISR   72
Ha   MRAAILLILFCSLQIHLTG~AK~IGHNAQKGIDDFLGGAHISKCGKHERILQGEDIIETNAKGELIEKIVSR   70
Ye   MRVAAFVILCCALQYV~~~~AANGYNDIQGVNTSNMRETDIEQTTDFEKDTNG~TLFEKTLTRKKFEGHLNV   67
Gm   MRVTTFVILCCALQYV~~~TADAIDDSLLNFNNENFIEIG~ESTTA~EVDVENGTLVERETTRKKYERDGDI   67
Bm   MRVKTFVILCCALQYVAYTNANINDFDEDYFGSDVTVQS~SNTT~~DEIIRDASGAVIEEQITTKKMQ~~~R   66
Ap   MRVIAFVILCCALQYA~~TAKNL~~~~~~~~~~~~~~~~~~~~RHHDEYVDNHGQLVERF~TRKHFER~~~N   47

Ro   REIITDDNSESESDSDSSEDSGSTEKIIKQIIIVQEKPKHGHH~~HAKEKIYEEEIIIKKIGDLPKKDCDE  141
Ha   KEILTDDDSESFSVSYSSEDD~STETIVKTITIVQEIPKHGKGKGHAKEKIFEEETVIKKIGKGVAAPVVA  140
Ye   GSG~PKLSGN~~~~~~~~~~~~~~~~DKIIRTFVIESDAS~~~~~~GQETIYEEDVVIHQVPGASASASAS  115
Gm   T~~~PNISGE~~~~~~~~~~~~~~~~DKIVRTFVIETDAS~~~~~~GHETVYEEDVVIKRKPGQQGVTERT  113
Bm   KNKNHGILGKN~~~~~~~~~~~~~~~EKMIKTFVITTDSD~~~~~~GNESIVEEDVLMKTLSDGTVAQSYV  116
Ap   AATRPHLSGNE~~~~~~~~~~~~~~~~RLVETIVLEEDPY~~~~~~GHEDIYEEDVVIKRVPGASSSAAAA   95
```

B

```
Ro   -70 SCGTIISDSSSNVGGLGRIGALHKLG~GGRIPRPAIYTRHPGLKGLKPQCPLSNFNVHVKVGNIRKANGNC
Ha   -70 VPGVISSGVVAPGLVGGHGGLVNVGGTH~VLPGTSVYTTHPDPRTVRSSCRTSPYNLLIKVGNARKLNGNC
Ld   -70 SSGSSSVESVRPGRVLAGLGRVGALGRSGVVPVPSVYTNHPGN~SVKSPCKLPDFNLFVKVGNVRKSNGNC
Ye   -68 GSAAASSAAAGSSGPVFVNNYGPYGSGSSAAASASASAGGYGSGGVRQPCYVSRRQLTVRVGSRRQ~CAFC
Gm   -69 APVVGPSVSSVGPVGAQVIEIGSPVVPSVSRTGSVSRVSVSGRPGVRVPCSLTRRQFVVKIGTRRQPCGYC
Bm   -67 GAGSGAGAGGSVSYGAGRGYGQGAGSAASSVSSASSRSYDYSRRNVRKNCGIPRRQLVVKF~~RALPCVNC
Ap   -54              VLDGGYDSEGSAAAAAAAAAAAASSSGRSTEGHPLLSICCRPCSHSHSYEASRISVH
```

Figure 2. H-fibroin nonrepetitive C-terminus (A) numbered from the translation-initiating Met (the signal sequence is shown in italics in Ap and N-terminus (B) numbered in reverse from the last amino acid residue in the caddisflies) *Rhyacophila obliterata* (Ro, unpublished), *Hydropsyche angustipennis* (Ha)[44] and *Limnephilus decipiens* (Ld)[44] and in the moths *Yponomeuta evonymella* (Ye),[26] *Galleria mellonella* (Gm),[38] *Bombyx mori* (Bm)[35] and *Antheraea pernyi* (Ap).[28] Small parts of the repetitive region are included (dash-underlined). Residues conserved at least in 3 species are highlighted grey, Cys residues in the secreted protein are printed boldface. A region conserved in the N-terminus of lepidopteran H-fibroins and in a very degenerate form in the trichopteran H-fibroins, is dot-underlined.

secretory unit.[21] Only a portion of the 30 kDa P25 and none of the less glycosylated 27 kDa form of P25 is present in the elementary secretory units,[6] but both forms occur in the spun-out silk. It is therefore not excluded that filament formation is associated with a partial deglycosylation or other changes in P25.

The conservation of L-fibroin, P25 and the nonrepetitive H-fibroin ends (Fig. 2) in the basal superfamily Yponomeutoidea,[26] suggests strongly that their interaction represents a plesiomorphic mechanism of the silk secretion and gel/filament conversion in Ditrysia. It was proposed that silk filament construction from the three protein types had been conserved because the hydrophobic nature of H-fibroin requires association with the L-fibroin and P25.[27] However, the H-fibroin of *Yponomeuta evonymella* is amphiphilic.[26] A different mechanism of the gel/filament conversion evolved in the family Saturniidae (Bombycoidea) that possesses relatively small H-fibroins (200-250 kDa) with alternating hydrophilic and hydrophobic motifs in their repeats.[28,29] This amphiphilicity probably allows gel formation and the gel/filament transition without the L-fibroin and the P25 that have been lost in this family.[30,31]

Filament polymerisation rests on weak molecular interactions between repetitive motifs that occupy about 95% of the H-fibroin length (Fig. 3). X-ray diffraction studies of lepidopteran silks revealed the presence of β-sheets whereby the carbonyl oxygen and amide nitrogen of adjacent peptide chains are linked by hydrogen bridges. Pleated β-sheets are stacked into crystallites held by interactions between amino acid side chains. From the crystallite dimensions it was predicted that they are largely based on Gly and Ala in *B. mori*[32] and on Ala in *Antheraea pernyi*.[33] This was later confirmed when the protein sequences became available. Most of the *B. mori* H-fibroin is built from the motif GAGAGS.[34] Strings of GAGAGS hexamers (S is occasionally replaced with Y or V) followed by GAAS or a similar tetrapeptide form a second tier repeat and two to six such repeats

Yponomeuta evonymella (Yonemura and Sehnal 2006)

```
B         GAAASSAAAAGSNDELVVNN YGGYGSGS
E         GAAASSAAAAGSSGPVVINNEYGPYGAGSSAAASAAAAGGAGGNGYGLNGYGA
E contd                         GGYGSGSGAGSSASAGGAGGYGSNGGAGGYGSGS
```

Galleria mellonella

```
A      GSSAASAASGASGAGPVIVIEDGSSAASAAAAGSGASGVGGLGLGGLGPLGGIGLIGASSASA
B1                                                 SGAGLGGVGAAGASGLGGLGGTGASAAGSAGAGLGGVGAGGSS
B2                                                                             TGSAAGSTGAGLGGSGAA
```

Bombyx mori

```
Reiterated motif:  (GAGAGS)11GAGAGAGYGAGYGAGAGAGYGAGAGSGAAS
„Spacer":                                     GAGAGAGAGAGTGSSGFGPYVANGGYSRSDGYEYAWSSDFGTGS
```

Antheraea pernyi

```
a        AAAAAAAAAAAAAGSGAGGSAGGGYGWGDGGYSDS
b            AAAAAAAAAAAAASGAGGAGGYGGYGGYGSDS
c             AAAAAAAAAAAAAGSGAGGRGDGDGGYGGYGSGSS
d                 AAAAAAAAAAAAARRGHDRAAGS
```

Hydropsyche angustipennis

```
HB       GPRGLGPLGLDSDIIGDGYGPYAL
HC                          GPRGLGPLGGLGRRPYGGYSASGSVSAEGPRGWYGPRGL
HC3                              GGIGGWRSRPYGGYSASHSVSVE
HD        APVVYHAPIIRRAPKISRSSSYSVERIVTPTVITRISGSHSVSAEGRRGVWGPHGVW
HA          GPSGLYGDSGIVDGVYGPGLVGPGGWGRRPYGGYSASRSVSAEGPRGWYGPRGL
```

Limnephilus decipiens

```
LD        ....VSISRSVSIERIVTPGVYTKISRSSSVSVEGGRRRGPWGYGRG
LE                       LSGSGDLDGLGGVGGLGGLGGLGGRRGPWGRGY
LF                                GSSGTVSVSVSVEEGRRRGPWGRRG
```

Figure 3. H-fibroin repeats in the moths *Y. evonymella*,[26] *G. mellonella*,[38] *B. mori*[35] and *A. pernyi*[28] and the caddisflies *H. angustipennis* and *L. decipiens*.[44] Several motifs are recognised: Ser and/or Ala rich, may alternate with Gly (dash-underlined), bulky hydrophobic central residues flanked with charged residues (highlighted grey), GlyGlyXxx triplets (dot-underlined) and motifs containing Pro (for example GPGLV) or Trp surrounded by charged residues (solid-underlined).

followed by a highly conserved sequence of 43 residues consisting of a chain of GA dipeptides followed by charged residues (Fig. 3) make up a third order iteration.[35] The 43 residue sequence functions as spacers breaking the repetitive region into 12 large modules. The repetitive H-fibroin region in *A. pernyi* also contains 12 modules, each consisting of 5-8 repeats. The repeats fall into four categories (Fig. 3) that are all composed of a relatively hydrophilic part (11-22 residues) and a hydrophobic string of 12-14 (exceptionally 5 and 15) alanines.[28] We wish to emphasize that the number or repeats per module varies; the same is true for the number of motifs concatenated in the repeats of *B. mori*. Such heterogeneity, occasional residue replacements and the intercalation of spacers separating the repeats probably prevent excessive crystallization that would occur with long strings of the precisely registered motifs.[36]

Bombyx and *Antheraea* represent families Bombycidae and Saturniidae that are characterised by low silk production in early larval instars and high production for the construction of extensive cocoons after the termination of feeding. During cocoon construction, up to 20% of the body biomass is converted to silk. Since sufficient endurance and strength are the major functional demands put on the cocoon silk, the use of simple crystalline motifs GAGAGS or poly(Ala) in the fibroin sequence is ideal. It is likely that these motifs are the products of selection for a high rate of silk synthesis at limited nutrient supply. Different selection forces operated in Pyraloidea and Yponomeutoidea that spin elastic hides during most of their larval life and eventually also construct cocoons for pupation. Their silk must be multifunctional and this is probably why their H-fibroins contain a variety of motifs. This variety presumably provides required combination of physical silk properties. Some motifs contain metabolically expensive amino acids. They are obtained from the food (most spinning is done during the feeding period) and from the old silk that is partly consumed and thereby being recycled.

Crystallites formed by stacked β-sheets were detected in the silks of all examined caterpillars[37] but differences in periodicity indicated that they are made by different amino acids. Analysis of H-fibroins in the representatives of Pyraloidea[38,39] and Yponomeutoidea[26] revealed that their repeats are formed by arrays of diverse motifs; motif concatenations are short and rare (Fig. 3). The Ala-rich motifs, which often consist of the strings of Ala or Ser, are likely to form crystallites; the replacement of Ala by Ser does not disturb crystallinity.[40] The formation of β-sheets and crystallites is also likely for the motifs in which Gly alternates with Ser or another residue, i.e., GXGX or GXXG. On the other hand, the GGX motif probably yields a spiral conformation.[41] Little is known about the conformation of the motifs that occur in the H-fibroin repeats of Pyraloidea and Yponomeutoidea (Fig. 3). These "amphiphilic" motifs consist of 3 hydrophobic amino acid residues flanked on each side by 2-4 polar residues. Motifs of this kind also occur in the nonrepetitive H-fibroin N-terminus in all Lepidoptera and in a more degenerate form in Trichoptera (Fig. 2). When present in the repeats, the motif is imbedded in a highly crystalline region and probably restrains the formation and size of the crystallites, thereby affecting fiber stiffness. Frequent association of this motif with Pro contributes to the idea that they function to disturb β-sheet formation.

The H-fibroins of Yponomeutoidea and Pyraloidea contain several repeat types (Fig. 3). Repeat composition from simple motifs as well as repeat arrangement in higher order modules is species specific. For example, *G. mellonella* H-fibroin includes highly conserved repeats of 63 amino acid residues (A), 43 residues (B_1) and 18 residues (B_2).[38] The repeat assembly $AB_1AB_1AB_1AB_2AB_2(AB_2)$ makes up a module and 12 such modules constitute the major part of the H-fibroin. High repeat homogeneity is probably essential for precise alignment of specific motifs that occupy precise but widely distributed regions of the repeats. The homogeneity is extreme in silks of high functional demands, such as in *G. mellonella*. Larvae of this species live in beehives and their survival depends on continuous spinning of tight tubes in which they hide. The tubes are elastic to allow the larva to turn around. On the other hand, stiffness and close fiber packing into a protective wall are needed in the cocoons that are spun at the end of larval development. The silk of *G. mellonella* probably meets all these requirements due to a proper blend of motifs and their precise alignment in the highly conserved repeats. Related species that are less dependent on the silk performance produce H-fibroins with less conserved repeats. In the Mediterranean flour moths, *Ephestia kuehniella*, the regular repeat arrangement is disturbed primarily in the central part of the repetitive region and in the Indian meal moth, *Plodia interpunctella*, the repeats seem to be erratic throughout the region.[39]

Caddisflies (Trichoptera)

The silk gland morphology and the silk composition in Trichoptera are similar to those of Lepidoptera. Histological evidence[42,43] and our unpublished cDNA sequences demonstrated the presence of sericin-like materials but none have yet been characterized. PSG proteins of the silk filament have been analysed in all three trichopteran suborders and found to be similar, in spite of differences in the silk use. The larvae of *Hydropsyche angustipennis* (suborder Annulipalpia, family Hydropsychidae) spin catching nets and hiding tubes and, at the end of larval development, construct small domes from the sand grains. The larvae of *Limnephilus decipiens* (Integripalpia, Limnephilidae) use silk to stick together pieces of plants and other materials (depending on the species) into protective cases; at the end of larval development they close the cases and pad them with silk lining. The larvae of *Rhyacophila obliterata* (Spicipalpia, Rhyacophilidae) use silk only before pupation to spin parchment-like cocoons. The silk of all three species contains homologues of lepidopteran H-fibroin and L-fibroin but not of P25.[44] Terminal regions of the H-fibroins and the entire L-fibroin sequences contain conserved spacing of residues with specific properties (hydrophobicity, charge), including the cysteines that link H-fibroin and L-fibroin in the silkworm silk. The structures of both the *H-fibroin* and *L-fibroin* genes are also similar to those of Lepidoptera, suggesting that the H-fibroin/L-fibroin interaction evolved in the ancestor of Trichoptera and Lepidoptera. P25 does not occur in Trichoptera but it is uncertain whether it has been secondarily lost in this order or if it represents an innovation specific to Lepidoptera-Ditrysia.

H-fibroin repeats in all analysed trichopteran H-fibroins are blends of several motifs that are rarely reiterated in string-like fashion (Fig. 3). Motif diversification is less pronounced than in Lepidoptera; the patterns of motif arrangement in the repeats, rather than differences in the motifs characterize caddisfly suborders. Existence of higher-order modules remains to be proven. The Ser-rich and Gly-rich motifs and an "amphiphilic" motif (2-3 hydrophobic residues flanked by 2 hydrophilic ones or a proline) resemble certain lepidopteran H-fibroins. The strings of SX dipeptides apparently form β-sheets and crystallites based on hydrogen bonding via polar zipper interactions,[45] whereas some of the Gly-rich strings, such as GGLGGLGH are more likely to have a spiral conformation. The GPGXX motif, which probably makes a β-spiral and confers filament elasticity,[46] occurs in the H-fibroin of *H. angustipennis* whereas species from other suborders contain Pro in other arrangements. Caddisfly H-fibroins do not contain the typical lepidopteran motifs GAGA and poly(Ala) and their Ala content is strikingly low with Ala virtually absent in the repeats of *L. decipiens*. Unique to caddisflies is a highly symmetrical motif with a central Trp (*R. obliterata*: GPGGRRGWGRRGPG). Its occurrence in all caddisfly suborders suggests functional importance. All examined caddisflies further contain repeats with an amphiphilic region that begins and ends with a $(\text{Ser-X})_{4\text{-}6}$ string and includes 30-35 amino acid residues.

Since the silks of caddisflies are produced and persist in water, one would expect H-fibroins with hydrophobic repeats but those identified so far are amphiphilic.[43] The high content of hydrophilic residues, both neutral and charged, probably facilitates the secretion and storage of the covalently linked L-fibroin/H-fibroin dimer in the absence of P25. Mechanism stabilising the spun-out silk in water remains to be discovered. It is not excluded that covalent cross-linking by disulfide bridges plays a role. The L-fibroins of caddisflies contain 2-5 additional Cys in comparison with the lepidopteran L-fibroins (Fig. 2). More importantly, several cDNAs from the silk gland-specific libraries of the caddisflies we studied encoded Cys-rich repeats (unpublished). A 37 kDa silk protein composed of 5 repeats, each 63-65 amino acids long and including 4-5 Cys residues, was discovered in another *Hydropsyche* species.[47]

Fleas (Siphonaptera)

Larvae of Siphonaptera (fleas) produce silk from the labial glands to construct nests, pupation cases and individual and group cocoons. There is no data on the silk composition, except for an over-all analysis of the representation of individual amino acid residues. The silk proteins have predominantly an α-helical conformation and the silk composition is characterised by low levels of glycine and a high proportion of glutamic acid residues.[2,37,48]

Midges, Glow Worms and Flies (Diptera)

Silk has evolved in the larvae of some nematoceran Diptera, which produce silk from the labial glands and independently in adult male hilarine flies (Diptera-Brachycera: Empididae). The male hilarine flies secrete class III silk from the dermal glands located in the fore leg basitarsi and use it to wrap gifts that are offered to females during courtship.[49] The silk contains a 220 kDa protein with unique repeats.[50] The silks of Nematocera have been described in the glow worm (Mycetophilidae; *Arachnocampa luminosa*) and analysed in detail in the water midges (Chironomidae). Glow worm larvae spin silken nests on the roof of caves or overhangs and suspend silken snares from this nest to trap prey. The silk proteins have not yet been described but X-ray diffraction data indicates that they are in a cross-beta structure.[51] Chironomid larvae spin protective tunnels throughout their development. Silk composition changes in the last larval instar when a pupation tube is made. The glands of *Chironomus tentans* include 38-40 cells producing about 15 secretory proteins that fall into three size groups: about 1000 kDa, 100-200 kDa and less than 100 kDa.[52] The largest ones (spIa, spIb, spIc and spId), are encoded by four closely related genes *BR1*, *BR2.1*, *BR2.2* and *BR6*, respectively; they are called *BR* to emphasize their localisation in the puffs (Balbiani rings) that are visible on the polytene chromosomes. These genes have unique 5' and 3' termini and a large central region consisting of uninterrupted blocks of 125-150 nearly identical repeats, each with a constant (C) and a slightly variable subrepeat (SR) region. The C region encodes about 30 amino acid residues with four Cys, one Met and one Phe in conserved spacing. The SR region codes for 30-60 residues that are arranged in shorter (3-11 residue) tandem repeats, rich in Pro, often flanked by basic and acidic amino acids (Fig. 4). Repeats encoded by *BR6* are extremely charged.

Seven silk proteins under 200 kDa are built from sequences related to the C and/or SR regions. For example, silk protein 115/140 includes 65-68 copies of an SR region, sp17 contains a single C/SR repeat and sp12 is a 7.6 kDa protein that may represent a diverged SR region.[53] The data suggest that 11 silk genes have evolved after sequence duplication and recombination of a single ancestral gene. The conservation of the C and SR sequences indicates the functional importance of their combination. Extreme deviation was found in the gene *BR3* that encodes one of the major silk proteins (designated sp185) in 38 exons (Fig. 4). The protein contains 72 copies of comparatively variable repeats with a C region. It has been proposed that splitting of the repetitive region into variable sized exons prevents the homogenization of the repeats observed in other *BR* silk genes. Structural studies on synthetic peptides representing the C or SR region indicated that the C region is primarily α-helical and the SR-region forms an extended poly(Gly) II-type helix.[54] It is assumed that the extended helical spI proteins assemble into a silk backbone by electrostatic interaction between the SR regions and covalent disulfide bridges between the C regions (there are >400 Cys per molecule). The propensity of spI silk proteins to form disulfide bridges was confirmed after a recombinant 83-mer protein containing typical C and SR regions was expressed in *Escherichia coli*.[55]

Chironomus thummi

BR1 gene, 210 bp repeat

C region: KCGSKMRRVLAEKCAARKGRFSASKCRCFS

SR region: RPSWSGIKPE KRSKSGSRPE KRSKSGSRPE KRSKSGSKPE

sp185/220

WDNNQCKCTCPVDMQMPVGGCGTGKS FNEFTCQCECEASASKCGLKR WSTDTCQCECPKPMPTEGCGPQS

Cotesia glomerata

28 amino acid repeat: NSGANANSNSNSNSNSNANSDANSNSYS

Apis melifera

AmelF1: RAGQV**A**LASQKD**A**VLQAQA**A**ASAASE**A**RAAAD

AmelF2: ELGAG**A**RAASVA**A**AAQAKN**T**EAAEAG**A**NAALA

AmelF3: VNAKA**A**VVKAS**A**LALAEA**Y**LRASAL**S**AAASA

AmelF4: GAGQS**A**AESQAL**A**ASQSKT**A**ANAAIG**A**SELTN

Figure 4. Some of the repeats identified in the silk proteins in the chironomid larvae and in the larvae of Hymenoptera. Repeats encoded by the BR genes of *Chironomus* contain a constant region C and the SR region composed of 4 subrepeats.[74] Smaller silk proteins, such as sp185/220 are composed of degenerate repeats.[75] The *Cotesia* silk protein contains the dipeptide repeat Asn (sometimes Gly, Glu or Tyr): Ser/Ala in a conserved 28 amino acid repeat.[62] The primary sequence of the coiled coil silk proteins (examples from *Apis melifera* are shown) have heptad repeats characteristic of coiled coil proteins in the seqeunce HPPHPPP where H = generally hydrophobic residues and P = generally polar or charged residues.[66] The initial residue of the heptad repeats is shown in boldface.

The silk glands of *Chironomus thummi* and related species have a lobe composed of 4 cells distinguishable from the others. These cells express a 2.5 kb transcript that encodes a putative 77 kDa protein with a comparatively nonrepetitive sequence unrelated to other chironomid silk proteins.[56] The sequence of 756 amino acid residues contains 24 Cys, 22 Met and strings of Ser and Thr. A secreted protein of 160 kDa is associated with both N- and O-linked glycan moieties. Some of these characteristics resemble lepidopteran sericins.

Sawflies, Ants and Bees (Hymenoptera)

Cocoon silks produced from the labial glands of Hymenoptera are based on a surprising array of protein structures. Below is a description of the phylogenetic distribution of the molecular structure of cocoon silk in six of the 21 hymenopteran superfamilies with the number of species from which the data was obtained indicated in brackets. The distribution of silk types between and among the two hymenopteran suborders Symphyta and Apocrita suggests that β-sheet structured silks are plesiomorphic, while coiled coil structured silks, polyglycine silks and collagen-like silks represent derived conditions.

Among Symphyta, production of silky cocoons is common. Silks of the superfamily Megalodontoidea (2 species) is β-sheet structured, similar to that of families Argidae (3 species), Cimbicidae (2 species), Diprionidae (3 species) and Pergidae (2 species) from the Tenthredionoidea.[2,37,57] Species from the family Tenthredinidae produce silk based on either β-sheet structured proteins, collagen-like proteins or polyglycine II structured proteins (4, 6 and 7 species, respectively);[2,58] with more than one of these protein structures detected in some of the species. Collagen is a structure where three left-handed polypeptide helices twist together to form a right-handed triple helix that is stabilized by numerous hydrogen bonds.[59] The interior of the triple helix cannot accommodate any side chains larger than hydrogen and this places a restriction on the primary sequence so that every third residue is required to be glycine. The collagen triple helix is typically stabilised by high levels of proline or 4-hydroxyproline (produced by post translational enzymatic hydroxylation of proline) in the nonglycine positions. As expected, the collagen from the Tenthredinidae cocoons contains around one third glycine but no hydroxyproline; instead it contains substantial amounts of hydroxylysine.[57] Polyglycine II is a structure of parallel poly-glycine chains, each with a three-fold screw axis and each hydrogen-bonded to six neighbors.[60] Consistent with the structural analysis, the polyglycine II silks of the sawflies contain high levels of glycine (66%).[2]

In the Apocrita, species from the parasitic wasp superfamilies Ichneumonoidea (5 species)[2,37,57] and Chrysoidea (2 species)[61] produce silk that is primarily based on β-sheet structured proteins. A partial gene sequence (785 bp) encoding a fragment of a very large silk protein (>500 kDa) has been described from the parasitic wasp *Cotesia glomerata* (Ichneumonoidea, Braconidae).[62] The partial sequence encodes a product that contains repeats of 84 amino acid residues (Fig. 4), including strings of the dipeptide $(N\text{-}S)_{14}$ (where Asn is substituted on one occasion for Gly, Asp or Tyr and Ser at four positions for Ala). Dipeptide repeats are characteristic of the β-sheet silk proteins but the presence of Asn as one of the residues is unusual. A comparison of 23 other Microgastrinae cocoon silks revealed that 19 had comparably high Asn levels (25-51%), whereas silks of other Braconidae was composed of predominantly Ala, Ser and Gly.[63] Where investigated, X-ray diffraction patterns indicate that the Braconidae silks are predominantly β-sheet structured.[2,37,57]

Larvae from the social superfamilies Apoidea and Vespoidea of the Apocrita produce silk from proteins in a tetrameric coiled coil structure.[57,64] Coiled coils are formed when multiple α-helical polypeptides wind around each other. The sequence of coiled coil proteins is characterized by a seven-residue periodicity where large, nonpolar residues generally occupy the first and fourth position of each heptad.[65] The coiled coil silk proteins have been described from five species: *Apis mellifera, Bombus terrestris* (Apoidea); *Oecophylla smargdina; Myrmecia forficata; Vespa simillima* (Vespoidea). In all species, four homologous genes were found.[61,66,67] They encoded small (30-50 kDa) proteins that consisted of a continuous predicted coiled coil region of around 210-residues (Fig. 4), flanked by shorter (20-180 residue) N- and C- termini. In the species from Vespoidea, the N- and C- terminals were rich in Ser. Despite the similarity in architecture, the proteins had diverged to less than 45% sequence identity between Apoidea and Vespoidea. This level of divergence was not unexpected as the character of the residues in the seven-residue periodicity of coiled coils is important rather than the amino acid identity. However, the composition of the seven-residue periodicity of the silks was unusual in comparison to classic coiled coils, containing very high levels of Ala in all positions and high levels of small polar residues (Ser and Thr) and low levels of large hydrophobic residues (Leu, Ile and Met) in the positions located within the core of the coiled coil.

All four of the above described silk protein structures are produced from the insects labial glands for the construction of cocoons that protect the metamorphosing juvenile. Are they possibly based on four different ancestral proteins the use of which was promoted in some and suppressed in the other taxa? Analysis of minor silk components may provide an answer to this question. Or did the genes encoding collagen, polyglycine or coiled coil silk proteins evolve from genes encoding β-sheet silk proteins? In all of the structures, space between and within the polypeptide chains is effectively filled generating structures with maximal van der Waals interactions and mechanically effective protein fibers. In order to pack effectively, there are constraints on the architecture and composition of the proteins that make up these filaments. Maximal van der Waals interactions will form in β-sheets when every second residue is consistent in size and bulkiness. Similarly, a protein is only able to form a polyglycine II lattice in the absence of nonglycine amino acids as there is no room for side chains larger than the hydrogen within the lattice. The structure of collagen is restricted to glycine in every third position in order for the requisite inter-chain hydrogen bonds to form. In comparison to the above structures, coiled coils are able to accommodate a wide range of structures, although glycine and proline are not favoured.

The co-incidence in the phylogenetic distribution of beta-sheet, polyglycine II, collagen-like structured silks in the Hymenoptera coupled with their common usage and method of synthesis, suggests an evolutionary link. However, it is difficult to envisage a gradual evolution of one protein structure to the other without intermediates with compromised mechanical integrity. Although the mechanical properties of most of the hymenopteran silks have not been described, it is known that in many cases the mechanical strength of silks appears to exceed that required for the silk function.[68] Therefore it is possible that the extraordinary mechanical properties of insect silks are incidental consequence of the silks tightly packed molecular structure. If the final structure offered other selective advantages then less mechanically efficient silk fiber intermediates may not be selected against. The lack of experimental comparison of the silks from each of these structures does not allow us to ascertain the selective advantage of either collagen-like or polyglycine II structured silk.

Although it is similarly difficult to trace an evolutionary path from the large, highly repetitive beta-sheet silk depicted by the *Cotesia* silk gene to the small nonrepetitive silks genes of the ants and bees, there is more information on the characteristics of the different silks that suggests why the coiled coil silks became so established. The coiled coil silks evolved in a common ancestor predating the emergence of sociality.[61] The coiled coil silks are more resistant to denaturation than the β-sheet silks and evolution of silks with enhanced stability would allow the construction of longer lasting domiciles.[61,66] Sociality is intrinsically linked to the ability to build durable nests and therefore evolution of sociality in the Apoidea and Vespoidea might be linked to the acquisition of this more stable silk.

Acknowledgements

Writing this chapter was in part supported by grant IAA5007402 from the Grant Agency of the Czech Academy of Sciences.

References

1. Sehnal F, Akai H. Insect silk glands: Their types, development and function and effects of environmental factors and morphogenetic hormones on them. Int J Insect Morphol Embryol 1990; 19:79-132.
2. Rudall KM, Kenchington W. Arthropod silks: The problem of fibrous proteins in animal tissues. Annu Rev Entomol 1971; 16:73-96.

3. Willis JH, Iconomidou VA, Smith RF et al. Cuticular proteins. In: Gilbert LI, Iatrou K, Gill S, eds. Comprehensive insect science, Oxford: Elsevier, 2005; 4:79-109.
4. Craig CL. Evolution of arthropod silks. Annu Rev Entomol 1997; 42:231-267.
5. Kristensen NP, Scoble MJ, Karsholt O. Lepidoptera phylogeny and systematics: The state of inventorying moth and butterfly diversity. Zootaxa 2007; 1668:699-747.
6. Inoue S, Tanaka K, Tanaka H et al. Assembly of the silk fibroin elementary unit in endoplasmic reticulum and a role of L-chain for protection of α1,2-mannose residues in N-linked oligosaccharide chains of fibrohexamerin/P25. Eur J Biochem 2004; 271:356-366.
7. Akai H, Nagashima T, Aoyagi S. Ultrastructure of posterior silk gland cells and liquid silk in Indian tasar silkworm, Antherae mylitta Drury (Lepidoptera • Saturniidae). Int J Insect Morphol Embryol 1993; 22:497-506.
8. Takasu Y, Yamada H, Tsubouchi K. Isolation of three main sericin components from the cocoon of the silkworm, Bombyx mori. Biosci Biotech Biochem 2002; 66:2715-2718.
9. Žurovec M, Yang C, Kodrík D et al. Identification of a novel type of silk protein and regulation of its expression. J Biol Chem 1998; 273:15423-15428.
10. Nirmala X, Mita K, Vanisree V et al. Identification of four small molecular mass proteins in the silk of Bombyx mori. Insect Mol Biol 2001; 10:437-445.
11. Gamo T, Inokuchi T, Laufer H. Polypeptides of fibroin and sericin secreted from the different sections of the silk gland in Bombyx mori. Insect Biochem 1977; 7:285-295.
12. Sprague K. The Bombyx mori silk proteins: characterization of large polypeptides. Biochemistry 1975; 14:925-931.
13. Grzelak K. Control of expression of silk protein genes. Comp Biochem Physiol 1995; 110:671-681.
14. Hamada Y, Yamashita O, Suzuki Y. Haemolymph control of sericin gene expression studied by organ transplation. Cell Differ 1987; 20:65-76.
15. Michaille JJ, Garel A, Prudhomme JC. The expression of five middle silk gland specific genes is territorially regulated during the larval development of Bombyx mori. Insect Biochem 1989; 19:19-27.
16. Garel A, Deleage G, Prudhomme JC. Structure and organization of the Bombyx mori Sericin 1 gene and of the Sericin 1 deduced from the sequence of the Ser 1B cDNA. Insect Biochem Mol Biol 1997; 27:469-477.
17. Huang J, Valluzzi R, Bini E et al. Cloning, expression and assembly of sericin-like protein. J Biol Chem 2003; 278:46117-46123.
18. Michaille JJ, Garel A, Prudhomme JC. Cloning and characterization of the highly polymorphic Ser2 gene of Bombyx-mori. Gene 1990; 86:177-184.
19. Michaille JJ, Garel A, Prudhomme JC. Expression of Ser1 and Ser2 genes in the middle silk gland of Bombyx mori during the fifth instar. Sericologia 1990; 30:49-60.
20. Žurovec M, Sehnal F, Scheller K et al. Silk gland specific cDNAs from Galleria mellonella L. Insect Biochem Mol Biol 1992; 22:55-67.
21. Inoue S, Tanaka K, Arisaka F et al. Silk fibroin of Bombyx mori is secreted, assembling a high molecular mass elementary unit consisting of H-chain, L-chain and P25, with a 6:6:1 molar ratio. J Biol Chem 2000; 275:40517-40528.
22. Hwang C, Sinskey AJ, Lodish HF. Oxidized redox state of glutathione in the endoplasmic reticulum. Science 1992; 257:1496-1502.
23. Tanaka T, Kajiyama N, Ishikura K et al. Determination of the site of disulfide linkage between heavy and light chains of silk fibroin produced by Bombyx mori. Biochim Biophys Acta 1999; 1432:92-103.
24. Takei F, Kikuchi Y, Kikuchi A et al. Further evidence for importance of the subunit combination of silk fibroin in its efficient secretion from the posterior silk gland cells. J Cell Biol 1987; 105:175-180.
25. Tanaka K, Inoue S, Mizuno S. Hydrophobic interaction of P25, containing Asn-linked oligosaccharide chains, with the H-L complex of silk fibroin produced by Bombyx mori. Insect Biochem Mol Biol 1999; 29:269-276.
26. Yonemura N, Sehnal F. The design of silk fiber composition in moths has been conserved for more than 150 million years. J Mol Evol 2006; 63:42-53.
27. Sehnal F, Žurovec M. Construction of silk fiber core in Lepidoptera. Biomacromolecules 2004; 5:666-674.
28. Sezutsu H, Yukuhiro K. Dynamic rearrangement within the Antheraea pernyi silk fibroin gene is associated with four types of repetitive units. J Mol Evol 2000; 51:329-338.
29. Hwang JS, Lee JS, Goo TW et al. Cloning of the fibroin gene from the oak silkworm, Antheraea yamamai and its complete sequence. Biotech Lett 2001; 23:1321-1326.
30. Tamura T, Inoue H, Suzuki Y. The fibroin genes of the Antheraea yamamai and Bombyx mori are different in the core regions but reveal a striking sequence similarity in their 5′-ends and 5′-flanking regions. Mol Gen Genet 1987; 206:189-195.

31. Tanaka K, Mizuno S. Homologues of fibroin L-chain and P25 of Bombyx mori are present in Dendrolimus spectabilis and Papilio xuthus but not detectable in Antheraea yamamai. Insect Biochem Mol Biol 2001; 31:665-677.
32. Marsh RE, Corey RB, Pauling L. An investigation of the structure silk fibroin. Biochem Biophys Acta 1955; 16:1-34.
33. Marsh RE, Corey RB, Pauling L. Structure of tussah silk fibroin. Acta Cryst 1955; 8:710-715.
34. Mita K, Ichimura S, James TC. Highly repetitive structure and organization of the silk fibroin gene. J Mol Evol 1994; 38:583-592.
35. Zhou CZ, Confalonieri F, Jacquet M et al. Silk fibroin: Structural implications of a remarkable amino acid sequence. Proteins Struct Func Genet 2001; 448:119-122.
36. Fedič R, Žurovec M, Sehnal F. The silk of Lepidoptera. J Insect Biotech Sericol 2002; 71:1-15.
37. Warwicker JO. Comparative studies of fibroins II. The crystal structures of various fibroins. J Mol Biol 1960; 2:350-362.
38. Žurovec M, Sehnal F. Unique molecular architecture of silk fibroin in the waxmoth, Galleria mellonella. J Biol Chem 2002; 277:22639-22647.
39. Fedič R, Žurovec M, Sehnal F. Correlation between fibroin amino acid sequence and physical silk properties. J Biol Chem 2003; 278:35255-35264.
40. Simmons A, Ray E, Jelinski L. Solid-state C-13 NMR of Nephila-clavipes dragline silk establishes structure and identity of crystalline regions. Macromolecules 1994; 27:5235-5237.
41. Ashida J, Ohgo K, Komatsu K et al. Determination of the torsion angles of alanine and glycine residues of model compounds of spider silk $(AGG)_{10}$ using solid-state NMR methods. J Biomol NMR 2003; 25:91-103.
42. Zaretschnaya SN. Glands of caddisworm. III. Spinning glands. Proc Acad Sci USSR 1965; 12:293-303.
43. Engster M. Studies on silk secretion in the Trichoptera (f. Limnephilidae). I. Histology, histochemistry and ultrastructure of the silk glands. J Morphol 1976; 150:183-211.
44. Yonemura N, Sehnal F, Mita K et al. Protein composition of silk filaments spun under water by caddisfly larvae. Biomacromolecules 2006; 7:3370-3378.
45. Bini E, Knight DP, Kaplan DL. Mapping domain structures in silks from insects and spiders related to protein assembly. J Mol Biol 2004; 335:27-40.
46. Hayashi CY, Lewis RV. Evidence from flagelliform silk cDNA for the structural basis of elasticity and modular nature of spider silks. J Mol Biol 1998; 275:773-784.
47. Eum JH, Yoe SM, Seo YR et al. Characterization of a novel repetitive secretory protein specifically expressed in the modified salivary gland of Hydropsyche sp (Trichoptera; Hydropsychidae). Insect Biochem Mol Biol 2005; 35:435-441.
48. Lucas F, Shaw JTB, Smith SG. Comparative studies of fibroins. I. The amino acid composition of various fibroins and its significance in relation to their crystal structure and taxonomy. J Mol Biol 1960; 2:339-349.
49. Yang JH, Merritt DJ. The ultrastructure of the silk-producing basitarsus in the Hilarini (Diptera: Empidinae). Arthrop Struct Dev 2003; 32:157-165.
50. Sutherland TD, Young JH, Sriskantha A et al. An independently evolved Dipteran silk with features common to Lepidopteran silks. Insect Biochem Mol Biol 2007; 37:1036-1043.
51. Rudall KM. Silk and other cocoon proteins. In: Florkin M, Mason HS, eds. Comparative Biochemistry, vol. 4. New York: Academic Press, 1962:397-443.
52. Wieslander L. The Balbiani ring multigene family—coding repetitive sequences and evolution of a tissue-specific cell-function. Progr Nucl Acid Res Mol Biol 1994; 48:275-313.
53. Galli J, Wieslander L. Structure of the smallest salivary-gland secretory protein gene. Chironomus-tentans. J Mol Evol 1994; 38:482-488.
54. Wellman SE, Hamodrakas SJ, Kamitsos EI et al. Secondary structure of synthetic peptides derived from the repeating unit of a giant secretory protein from Chironomus tentans. Biochim Biophys Acta 1992; 1121:279-285.
55. Smith SV, Correia JJ, Case ST. Disulfide bonds in a recombinant protein modeled after a core repeat in an aquatic insects silk protein. Protein Sci 1995; 4:945-954.
56. Hoffman RT, Schmidt ER, Case ST. A cell-specific glycosylated silk protein from Chironomus thummi salivary glands—Cloning, chromosomal localization and characterization of cDNA. J Biol Chem 1996; 271:9809-9815.
57. Lucas F, Rudall KM. Extracellular fibrous proteins: The silks. In: Florkin M, Stotz EH, eds. Comprehensive Biochemistry. Amsterdam: Elsevier, 1968; 26B:475-558.
58. Lucas F, Rudall KM. Variety in composition and structure of silk fibroins: Some new types of silk from the Hymenoptera. Symposium on Fibrous Proteins. Canberra: Butterworths 1967:45-55.
59. Ramachandran GN, Kartha G. Structure of collagen. Nature 1955; 176:593-5.
60. Crick FHC, Rich A. Structure of polyglycine II. Nature 1955; 176:780-781.

61. Sutherland TD, Weisman S, Trueman HE et al. Conservation of essential design features in coiled coil silks. Mol Biol Evol 2007; 24:2424-2432.
62. Yamada, H, Shigasada K, Igarashi Y et al. A novel asparagine-rich fibrous proteins (Xenofibron) from the cocoons of the parasitic wasp Cotesia (=Apanteles) glomerata. Int J Wild Silkmoth and Silk 2004; 9:61-66.
63. Quicke D, Shaw M, Takahashi M et al. Cocoon silk chemistry of noncyclostome braconidae, with remarks on phylogenetic relationships within the Microgastrinae (Hymenoptera: Braconidae). J Natural History 2004; 38:2167-2181.
64. Atkins EDT. A four-strand coiled coil model for some insect fibrous proteins. J Mol Biol 1967; 24:139-141.
65. Woolfson DN. The design of coiled coil structures and assemblies. In: Parry DAD, Squire JM, eds. Fibrous Proteins: Coiled-coils, Collagen and Elastomers. San Diego: Elsevier, 2005:79-112.
66. Sutherland TD, Campbell PM, Weisman S et al. A highly divergent gene cluster in honeybees encodes a novel silk family. Genome Res 2006; 16:1414-1421.
67. Sezutsu H, Kajiwara H, Kojima K et al. Identification of four major hornet silk genes with a complex of alanine-rich and serine-rich sequences in Vespa simillima xanthoptera Cameron. Biosci Biotech Biochem 2007; 71:2725-2734.
68. Denny MW. Silks—Their properties and functions. In: Vincent JFV, Currey JD, eds. The Mechanical Properties of Biological Materials. Cambridge: University press Cambridge, 1980; 34:247-272.
69. Rentz DCF. The world's most unusual gryllacridid (Orthoptera: Gryllacrididae). J Orthoptera Res 1997; 6:57-68.
70. Morris DC, Schwarz MP, Cooper SJB et al. Phylogenetics of Australian Acacia thrips: The evolution of behaviour and ecology. Mol Phylogen Evol 2002; 25:278-292.
71. Fletcher MJ, Kent DS. Feeding by kahaono leafhoppers in silken shelters (Hemiptera: Cicadellidae: Typhlocybinae: Dikraneurini). Austral Entomol 2002; 29:115-118.
72. Paulsson G, Höög C, Bernholm K et al. Balbiani ring 1 gene in Chironomus tentans: Sequence organization and dynamics of a coding minisatellite. J Mol Biol 1992; 225:349-361.
73. Case ST, Cox C, Bell WC et al. Extraordinary conservation of cysteines among homologous Chironomus silk proteins sp185 and sp220. J Mol Evol 1997; 44:452-462.
74. Höög C, Wieslander L. Different evolutionary behavior of structurally related, repetitive sequences occurring in the same Balbiani ring gene in Chironomus tentans. Proc Natl Acad Sci USA 1984; 81:5165-5169
75. Takasu Y, Yamada H, Tamura T et al. Identification and characterization of a novel sericin gene expressed in the anterior middle silk gland of the silkworm Bombyx mori. Insect Bioch Mol Biol 2007; 37:1234-240

CHAPTER 9

The Elaborate Structure of Spider Silk: Structure and Function of a Natural High Performance Fiber

Lin Römer and Thomas Scheibel*

Abstract

Biomaterials, having evolved over millions of years, often exceed man-made materials in their properties. Spider silk is one outstanding fibrous biomaterial which consists almost entirely of large proteins. Silk fibers have tensile strengths comparable to steel and some silks are nearly as elastic as rubber on a weight to weight basis. In combining these two properties, silks reveal a toughness that is two to three times that of synthetic fibers like Nylon or Kevlar. Spider silk is also antimicrobial, hypoallergenic and completely biodegradable.

This article focuses on the structure-function relationship of the characterized highly repetitive spider silk spidroins and their conformational conversion from solution into fibers. Such knowedge is of crucial importance to understanding the intrinsic properties of spider silk and to get insight into the sophisticated assembly processes of silk proteins. This review further outlines recent progress in recombinant production of spider silk proteins and their assembly into distinct polymer materials as a basis for novel products.

Preface

Spider silk has attracted human interest for thousands of years,[1-3] mostly due to its toughness and ductility, but also because spider silk seems not to cause inflammation and allergic reactions. Therefore spider silk has been employed for hunting and fishing as well as for bandages.[1]

In nature, spiders use their silk for several applications such as for webs, wrapping of prey, protection of their offspring and as a lifeline which ensures their safe escape from predators (Fig. 1A). The variety of uses for silk among spiders is much larger than among insects like silkworms (see chapter 8), which use their silk often for making cocoons.

Catching Prey in Mid-Flight

Currently, more than 34.000 different spider species are known and roughly 50% thereof use webs to catch prey. Further, more than 130 different shapes of spider webs are known. Among the most studied webs are the so-called orb webs which consist of several different types of silk as shown for the European garden spider *Araneus diadematus* (Fig. 1A, Fig. 1B).[4]

Frame and radii of orb webs are made of strong and rather rigid silk. The underlying proteins (typically two different types) are produced in the major ampullate glands and therefore silk fibers made of these proteins are named MA silk. This particular silk is further used by spiders as a lifeline (or roping thread) which has to be ready at hand to escape predators—it is therefore always dragged, hence the nickname "dragline silk". The capture spiral of an orb web comprises fibers of

*Corresponding Author: Thomas Scheibel—Universität Bayreuth, Fakultät für angew. Naturwissenschaften, Lehrstuhl für Biomaterialien, Universitätsstr. 30, 95447 Bayreuth, Germany. Email: thomas.scheibel@uni-bayreuth.de

Fibrous Proteins, edited by Thomas Scheibel.

only one type of protein which is produced in the flagelliform (Flag) gland of spiders. Flag silk is highly elastic (up to 300%) and perfectly dissipates the impact energy of prey. For example, a typical honey bee with a body weight of 120 mg and a maximum flight velocity of about 3.1 m/s crashes into a spider's web with a kinetic energy of approx. 0.55 mJ.[5] Flag silk with diameters of only 1-5 μm can sufficiently withstand that (on this scale) massive impact. The enormous resilience of these threads is crucial for catching and holding prey which is sometimes even bigger than the spider itself.

In addition to MA and Flag silk, orb weaving spiders utilize two further silks for constructing webs. Silk fibers made of proteins produced in the minor ampullate gland are used to build an auxiliary spiral. This temporary spiral stabilizes the body of the web and provides a template for the capture spiral.[6] Connections between Flag and MA silk (the web scaffolding connection joints)

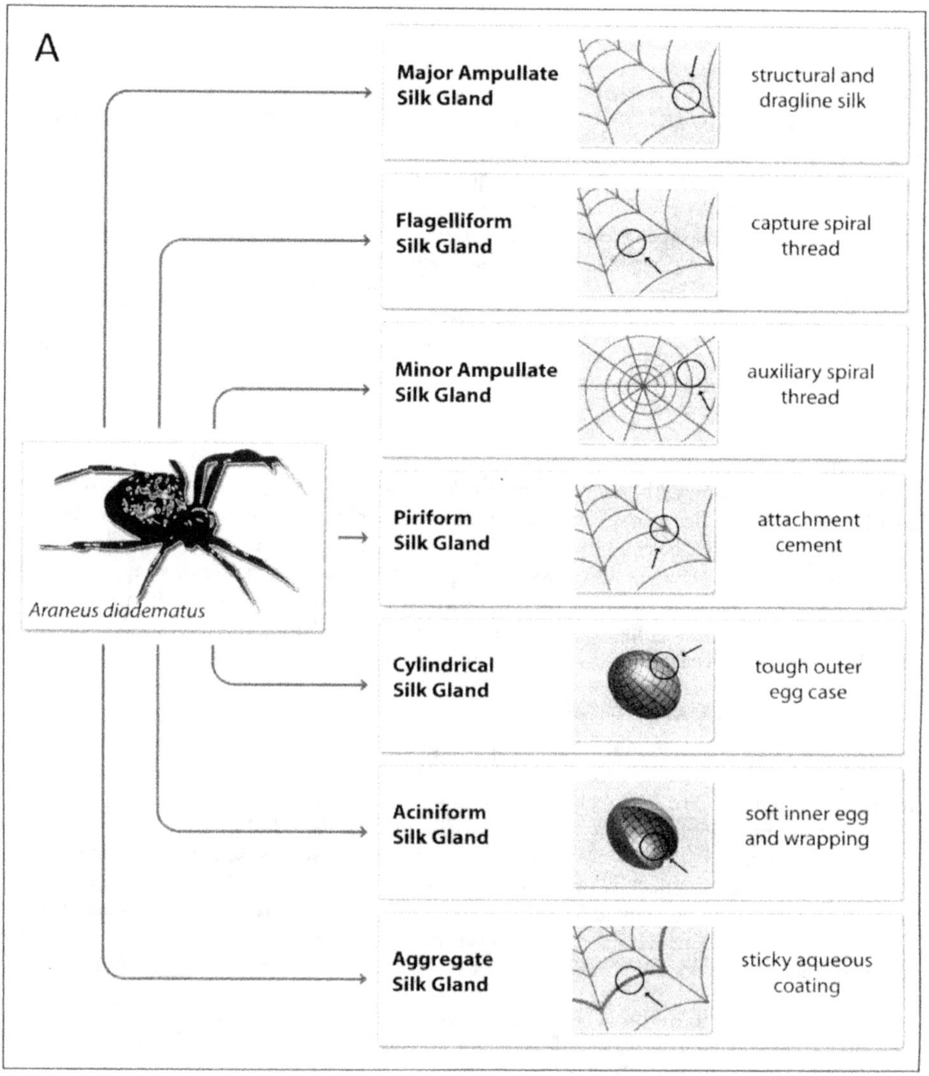

Figure 1. Legend continued on following page.

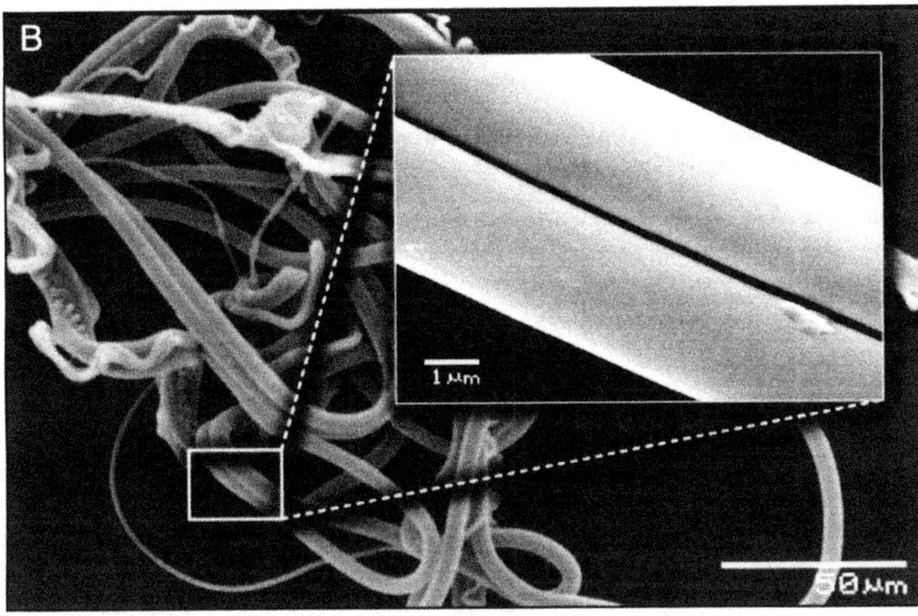

Figure 1, continued from previous page. A) Schematic overview of different types of spider silk as produced by female orb weaving spiders such as the European garden spider *Araneus diadematus*. B) Electronmicrographs of natural spider silk taken from an orb web of *Araneus diadematus*. The various morphologies of the distinct silk types can be easily depicted. The inset shows the smooth surface of MA silk.

and attachment of the web frame to the substratum (trees, undergrowth, masonry,...) are made of a sophisticated silk cement consisting of proteins produced in the piriform gland.[7]

Importantly, Flag silk is not sticky. To retain prey in the web, additional glue is applied around the capture threads.[8] Evolutionarily, the first adhesive was a special silk fiber which was hackled by comb-like devices, cribella, on the spider's hind legs in order to achieve a maximum surface area. The adhesive properties of cribellate silk are mainly based on a large sum of van-der-Waals forces provided by huge surface areas similar to the adhesion principle of gecko feet, where an extremely large surface area is provided by nano hairs on the toes.[9,10] Although an excellent adhesive, hackling of silk consumes a huge amount of time and energy. Thus, another type of spiders, ecribellate spiders, developed a different strategy of gluing. Specific silk threads are covered with a sticky aqueous layer containing organic molecules, salts, fatty acids and small glycoproteins which are produced in the aggregate gland of spiders.[11,12] Recently, it was proposed that the wet glue also contains small peptides which are thought to function as metal chelators.[8] The presence of metal ions might contribute to the inhibition of microbial growth on silk threads which is crucial to protecting the web or preventing damage to developing eggs. Wet glue is more energy- and time-efficient (compared to the cribellate system) and is therefore used by the majority of today's spiders.

The Structure of Spider Silk

Primary Structure

Spider silk primarily consists of proteins that possess large quantities of nonpolar and hydrophobic amino acids like glycine or alanine, but for example, no or only very little tryptophan.[4,13,14] In comparison to common cellular enzymes, it is evident that silk proteins exhibit a quite aberrant amino acid composition (Fig. 2A). Furthermore, spider silk proteins contain highly repetitive amino acid sequences, especially in their large core domain (Fig. 2B).

The repetitive sequences often account for more than 90% of the whole spider silk protein and are composed of short polypeptide stretches of about 10-50 amino acids. These motifs can be repeated more than a hundred times within one individual protein. Each polypeptide repeat therefore has distinct functional features resulting in the outstanding mechanical properties of spider silk threads.[15] MA and Flag silks contain up to four typical oligopeptide motifs which are repeated several times: [I] $(GA)_n/(A)_n$, [II] GPGGX/GPGQQ, [III] GGX (X = A, S or Y) and [IV] "spacer" sequences which contain charged amino acids (Fig. 2C).[16,17] Structural analysis revealed that oligopeptides with the sequence $(GA)_n/(A)_n$ tend to form α-helices in solution

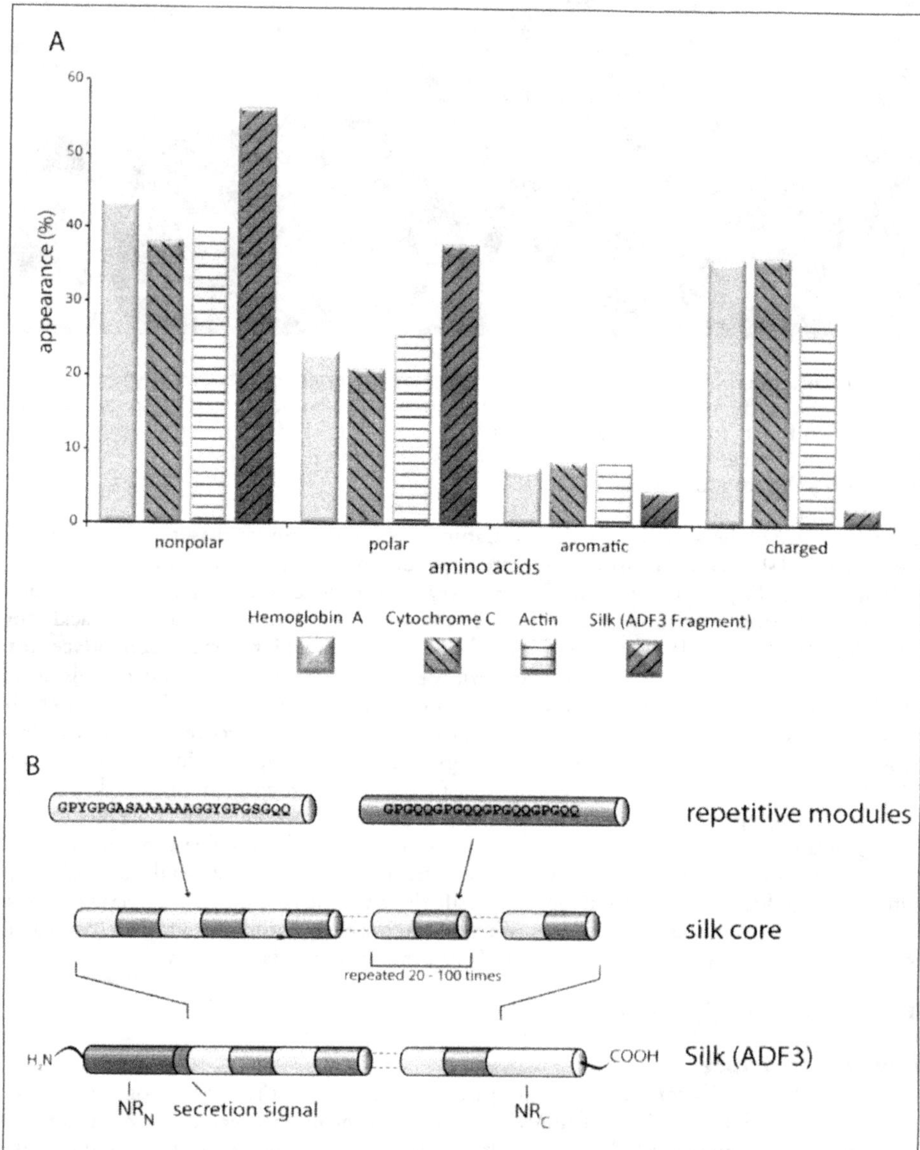

Figure 2. Figure legend viewed on following page.

C

	protein origin	amino acid sequence	putative structure	property
motifs	MA	$(GA)_n$ / A_n	β-sheet	strong
	MA	(GPGQQ)	β-spiral	elastic
	Flag	(GPGGX)		
	MA/Flag	(GGX)	3_{10}helix	elastic
	Flag	spacer	unstructured	elastic
termini	MA/Flag	NR_N-domain	helical	signal sequence and unknown
	MA/Flag	NR_C-domain	helical	assembly

Figure 2, continued from previous page. A) Amino acid composition of three common proteins in comparison to spider silk. B) Model of the hierarchical structure of a MA silk protein. For example, *Araneus diadematus* Fibroin-3 (ADF3) has a highly repetitive core domain flanked by two nonrepetitive domains (aminoterminal domain: NR_N; carboxyterminal domain: NR_C). The aminoterminal domain also comprises a secretion signal sequence to allow protein export. C) Amino acid motifs of silk proteins. The different amino acid motifs present in MA and Flag silk are correlated with their putative structure and their impact on the final properties of the thread. Although the function of the nonrepetitive parts is not fully understood, it is thought that these parts play a role in triggering the assembly of spider silks.

and β-sheet structures in assembled fibers.[18,19] The structures acquired by oligopeptides with the sequences GPGGX/GPGQQ and GGX have not yet been identified. Several studies describe these regions to adopt amorphous rubber-like structures,[20,21] whereas others suggest formation of a 3_1-helical structure.[18] Flagelliform silk, typically rich in GPGGX and GGX motifs, preferably folds into β-turn structures resulting in a right-handed β-spiral helix upon stacking (13,14,30).[16,22]

Apart from the repetitive core domain, nonrepetitive regions are located at the protein's termini.[23] These nonrepetitive terminal domains of the proteins are crucial for the assembly of spider silk proteins into fibers. The regions comprise approx. 100-200 amino acids and show—in contrast to the repetitive core—well defined secondary and tertiary structures in solution.[24,25] Due to conserved cysteine residues, these domains can establish intermolecular disulfide bonds and are thus able to stabilize dimers and multimers under oxidizing conditions. Therefore, these domains are thought to initiate and specify assembly of silk proteins.[26-28] Several carboxy-terminal nonrepetitive sequences of different silks and spiders have been identified, revealing a high sequence homology amongst these domains.[24,29,30]

However, only one full length sequence of a Flag silk protein from *Nephila clavipes* covering both termini has been identified so far. Further, only recently the first full length sequence of MA silk from the black widow spider has been reported.[31]

In contrast to the Flag silk gene, both analyzed MA silk genes lack introns consisting instead of unusually large exons (>9000 bp of coding sequence). The large exons might be a result of gene duplication processes during evolution. Additionally, genes with shorter introns generally show a higher expression rate than genes with large introns[32]—and spider silks are highly expressed throughout the lifetime of the spider.

The primary structure of spider silk proteins shows a specific hydrophobicity pattern with alternating hydrophilic and hydrophobic blocks in their core domains. Such amphiphilic composition is reminiscent of surfactants or biological membranes and, in the case of spider silks, is thought to be crucial for phase separation during the spinning process (see below).[33-35] Additionally, the unusual amphiphilic pattern might be responsible for formation of micelles postulated as intermediate structures during thread assembly (see below).[36]

Quaternary Structure and Protein Stability

After secretion from the silk glands, silk proteins are in aqueous solution and lack considerable secondary or tertiary structure.[37] Particularly in their repetitive core domains, however, the long repetitive sequences permit weak but numerous intra- and intermolecular interactions between neighboring domains and proteins upon passage through the spinning duct. These interactions result in the formation of secondary, tertiary and quaternary structure. Roentgen diffraction analysis of the final structure of MA silk threads led to the identification of areas of high electron density embedded in areas with low electron density (Fig. 3).[2,3,38] In a postulated model of this structure, the high electron density regions comprise crystalline sub-structures with high β-sheet content.[39] These sub-structures are thought to be responsible for the mechanical strength of the silk thread.

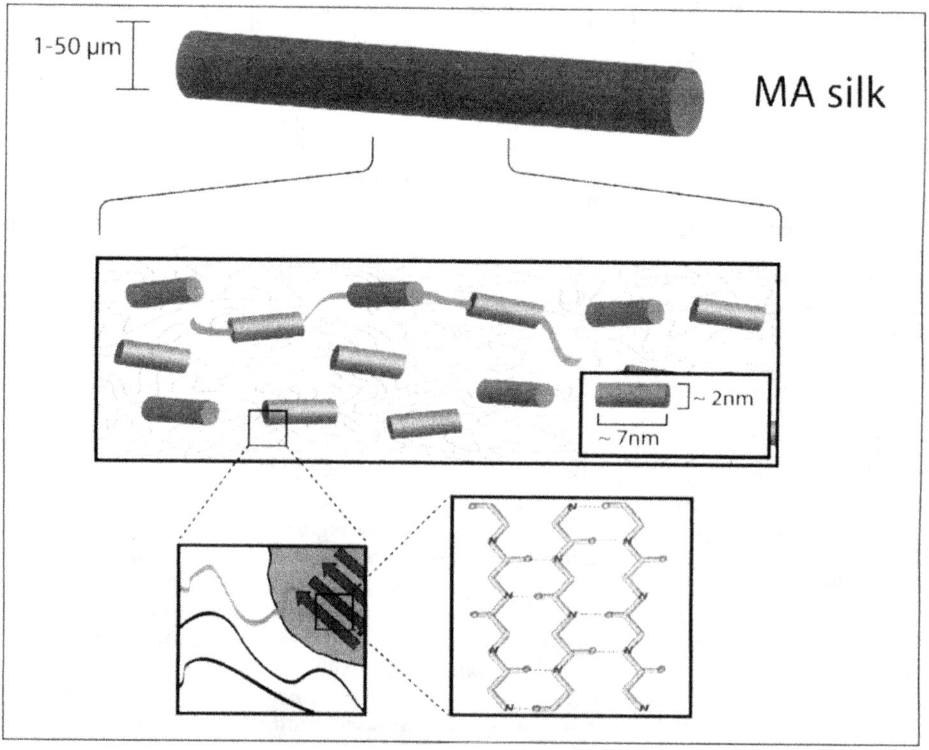

Figure 3. Schematic structure of spider MA silk. The thread is composed of small crystalline β-sheet rich subunits (see close-ups) which are embedded into an amorphous structure. The crystalline and noncrystalline parts are covalently connected, ensuring the coexistence of strength and ductility. Diameters of MA threads depend on species, but also on age, weight and state of health of a specific individual.

The elasticity of silk is based on the areas with low electron density, which are characterized by amorphous structures with few defined elements of secondary or supersecondary structure.[40,41] Such arrangement closely resembles that of protein hydrogels.[42] Upon tensile loading, the hydrogel-like areas can partially deform, contributing to the elasticity and flexibility of the thread.

Different types of silk reveal different structural distributions (e.g., different compositions of crystalline- and hydrogel-parts). MA silk which is used for constructing the frame of the web contains a high amount of crystalline (β-sheet) structures. In contrast, the much more flexible Flag silk consists almost exclusively of amorphous hydrogel-like regions. Thus, the correlation between the structure and function of individual spider silk proteins becomes evident. However, in the future more detailed analysis is necessary to characterize the structure-function relationship of individual spider silk proteins.

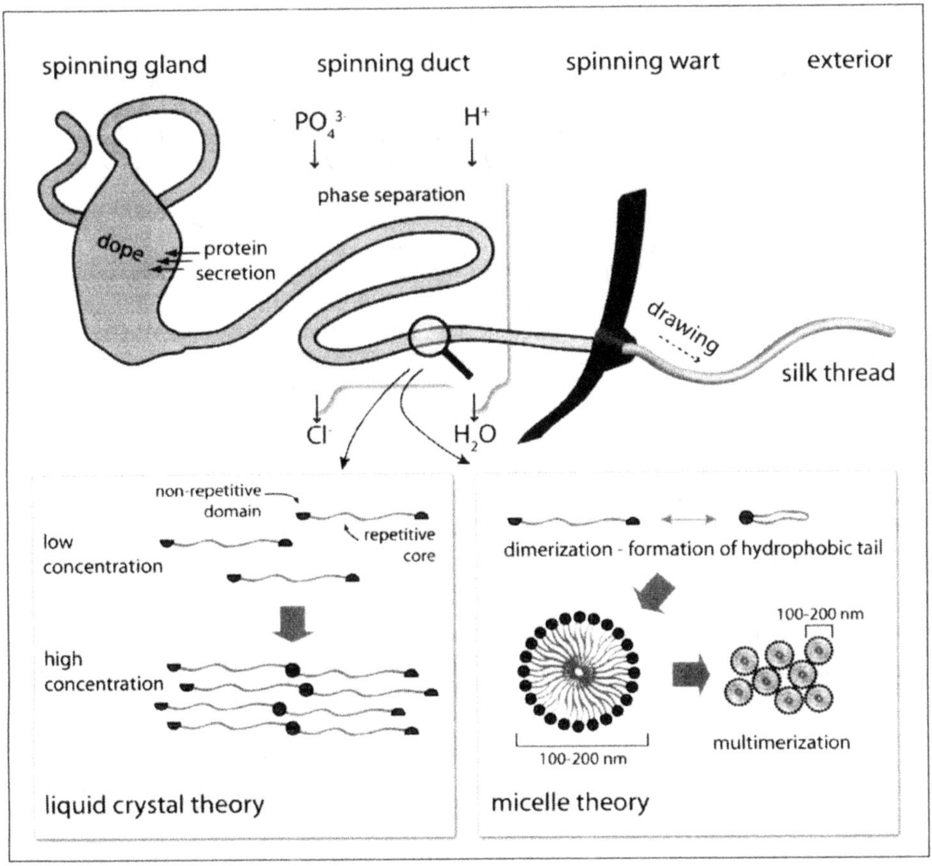

Figure 4. Model of the silk spinning process. The highly concentrated spider silk protein solution is secreted and stored inside the spinning gland. Upon initiation of thread formation, the solution is directed through a narrow ion exchange channel, in which a phase separation process takes place. Mechanical drawing of the silk thread (hind legs/gravity) results in rapid assembly of the silk fiber. Two theories exist concerning the molecular process inside the spinning duct. The first theory supports a liquid-crystalline behavior of the silk proteins, whereas the other theory favors a micellar organization of the proteins, before elongation (through laminar force) finally leads to thread assembly.

The Assembly of Spider Silk

Starting with a Highly Concentrated Spinning Dope...

As described above, assembly does not commence with globular folded protein monomers, but with mainly intrinsically unfolded proteins at extremely high concentrations.[43] Several mechanisms are thought to be necessary to gain and maintain the high silk protein concentrations (up to 50% w/v)[44] in the gland, including lyotropic liquid crystallinity, glycosylation of the outer surface of the folded silk proteins and phase separation induced by a polyol or by a phospholipid surfactant.[45] Starting with an almost entire random coil structure in the gland, silk proteins rapidly assemble upon passage through the spinning duct and the silk structure becomes water-insoluble (Fig. 4).[46] The tightly controlled assembly behavior requires bi-stable folding of the involved protein and precise control of the environmental conditions in the spinning duct (e.g., pH, ionic concentration, water content).

Phase Transition in the Spinning Duct: Two Theories

Successful silk assembly is based on extending, aligning and packaging of individual silk proteins in the laminar flow inside the spinning duct. Hydrophobic residues (typically polyalanines of the repetitive core region) align upon multimerization initiated by the terminal domains and additionally by shear forces in the spinning duct. These polyalanine segments thereby expose an increasingly hydrophobic surface which triggers the formation of β-pleated structures with numerous intra- and interchain hydrogen bonds.[46]

Two theories on the mechanism of silk fiber assembly have been proposed.[36,37,47] One is based on the crystalline alignment of the underlying proteins in the laminar flow inside the spinning duct (Fig. 4). Monomers or disulfide-linked multimers pass the spinning duct at very high concentrations. The alignment in one direction together with the high concentration results in a liquid-crystalline like behavior of the spinning dope. The proposed liquid crystalline state is the basis for the formation of intermolecular interactions like van-der-Waals forces and hydrogen bonds between neighboring molecules.[37] Upon further loss of solvent the conformational conversion is finalized and a silk fiber can be drawn out of the spinning duct.

In the second model, silk proteins first assemble into small micelles with a diameter of approx. 100-200 nm due to their amphiphilic properties inside the spinning dope.[36] A multitude of these micelles form globules with diameters in the micrometer range (Fig. 4). Shear forces, which arise during passage through the spinning duct, force these globules into an elongated shape finally leading to fiber formation.

Table 1. Mechanical properties of natural and synthetic fibers (taken from references 34 and 47)

Material	Density [g cm^{-3}]	Strength [Gpa]	Elasticity [%]	Toughness [MJ m^{-3}]
MA silk*	1.3	1.1	27	180
Flag silk*	1.3	0.5	270	150
Insect silk#	1.3	0.6	18	70
Nylon 6.6	1.1	0.95	18	80
Kevlar 49	1.4	3.6	2.7	50
Carbon fiber	1.8	4	1.3	25
Steel	7.8	1.5	0.8	6

*European garden spider *Araneus diadematus*
#Silkworm *Bombyx mori*

Mechanical Properties of Spider Silk

The most outstanding property of spider silk is its maximal resilience. Distinct spider silk threads are able to absorb three times more energy than for example Kevlar, one of the sturdiest materials on a weight-to-weight basis (Table 1).[39] It is interesting to note that synthetic materials typically show a higher stiffness and strength compared to natural fibers, whereas natural fibers tend to be more elastic. Synthetic carbon fibers, for example, have a yield point at approx. 4 GPa. This is more than five times higher than the best insect silk. The elasticity of carbon fibers, on the other hand, is only marginal. As soon as a carbon fiber is elongated more than 1 percent or bent to a certain degree it will instantly break. Spider silk shows a well-balanced combination of strength and elasticity and therefore mechanically outperforms other natural fibers as well as synthetic threads under certain circumstances.[48]

In addition to its outstanding resilience, MA silk shows a torsional shape memory that prevents the spider from twisting and turning during its descent on a MA silk thread.[49,50] Interestingly, MA silk needs no extra stimulus for total recovery after being turned from its initial position. Instead, it scarcely oscillates after twisting because of its high damping coefficient. Spider silk also shows a high supercontraction rate.[51,52] Absorption of water leads to shrinkage and tightens the thread. This process is important to ensure the rigidity of the spider's web during its lifetime and is thought to be caused by the organization and arrangement of individual silk proteins.[53,54]

Spider Silk vs Insect Silk

Spider silk is often compared to insect silk, preferably taken from the silkworm *Bombyx mori*. The commercially available silkworm silk is reeled from cocoons of caterpillar pupae. This process has only been slightly optimized over centuries and is highly cost-efficient. MA spider silk can be obtained by manually drawing the silk thread out of the spinning wart of immobilized spiders. However, this process is only suitable for MA silk (and not for the other spider silks), it is time consuming and highly expensive, especially since most spiders are cannibals, rendering farming costly.

The differences between insect and spider silks are evident on all levels, from the molecules involved to the structural arrangement of the proteins to the mechanical properties of the thread. On a molecular level, insect silk comprises a large amount of sericin proteins, which are absent in spider silk. The proteins which are responsible for the fibrillar structure (so-called fibroins in insect silk) are, in contrast to spider silk spidroins, composed of light and heavy chain counterparts. Mechanically, silkworm silk is much weaker and less extensible as compared to for example MA silk of spiders.[55,56] Interestingly, depending on spinning conditions, silkworm silk is either strong or elastic, whereas spider silk combines both properties.[57]

Although the mechanical properties of both types of silk crucially depend on spinning conditions, it is primarily the proteins involved that make the real difference. Therefore, techniques have long been sought to recombinantly produce and engineer natural spider silk proteins.

Mimicking Nature

Recombinant Spider Silk

Several biotechnological methods for recombinantly producing spider silk proteins have been analyzed during the last decades, since recombinant production of sufficient amounts of silk proteins is essential for understanding their structure and their assembly behavior.[48] Due to the highly repetitive character of individual spider silk molecules it is quite complicated to determine the complete cDNA sequence of a silk gene. As mentioned above only limited information about complete silk genes is available. Sequencing of repetitive proteins often gives incorrect or unreliable results, providing little information on gene size or the number of repeat units in particular.

The first attempt to obtain recombinant spider silk proteins was the direct transformation of original or fragmented silk genes into bacterial hosts. Bacteria, however, were not suitable for this task due to the large size of the genes..[58] Recombinant production of spider silk proteins in bacteria has been further complicated by the different codon usage of spiders compared to bacteria.[14]

Thus, eukaryotic expression systems like the yeast *Pichia pastoris* were tested as expression hosts.[59] Here, problems occurred during protein purification rather than during the production process. Similar problems were encountered upon attempts to use plants (such as potato or tobacco) as transgenic expression host, however, such systems are still under consideration since they appear to be attractive for larger scale productions.[60]

Other expression systems for the direct transformation of spider silk genes/fragments have also been used during the last decades. The most prominent one is the expression of spider silk in the milk glands of transgenic goats by the Canadian company Nexia Biotechnologies. Although this technique initially produced promising results, the concentration of soluble protein in the milk was found to be low and the proteins could not be efficiently purified for thorough analysis.[61] Tests were also performed in mammalian cell lines with comparable results.[62]

Finally, insect cells which are phylogeneticaly closely related to that of spiders were employed using the baculovirus expression system.[63,64] This particular system benefits from the fact that the easily modifiable baculovirus selectively infects insect cells. Partial cDNAs comprising the known carboxyterminal repetitive and nonrepetitive sequences of the two MA proteins of *Araneus diadematus*, ADF3 and ADF4, were cloned into the baculoviral genome. The recombinant virus was then used to infect insect cell lines (e.g., sf9 cells of the fall armyworm *Spodoptera frugiperda* or high five cells of the cabbage looper *Trichopulsia ni*). During cytoplasmic production, spider silk either remained soluble or assembled into solid fibers—depending on the type of MA silk protein.[48,64] A great advantage of the baculovirus expression system, in comparison to expression in mammalian cells, is the relatively easy culturing process and the potential to easily modify the expression conditions.[48] For large scale production, however, the Baculovirus system might be too inefficient and too expensive.

The difficulties encountered made it evident that a direct transformation of original silk genes and silk fragments is not the method of choice for recombinant spider silk protein production. To achieve stable production with efficient yields, i.e., by bacterial hosts, the spider genes must be adapted to the bacterial codon usage. To this end, genes coding for spider silk-like proteins were generated using a cloning strategy which is based on a combination of synthetic DNA modules and PCR-amplified authentic gene sequences.[28] This approach includes the carboxyterminal nonrepetitive regions present in all MA silks and the codon-optimized repetitive regions. Using a seamless cloning technique, a controlled combination of engineered modules as well as of authentic gene fragments without generation of unwanted gaps or artifacts inside the coding sequence is possible.[65] Employing this strategy, it was possible to dramatically increase the expression yields.[66] The modular system further enabled the production of special engineered proteins with tailor-made properties for experimental analysis.

Artificial Spinning of Spider Silk

Due to the availability of recombinant spider silk proteins, scientists will be able to analyze the assembly of spider silk threads in a functional in vitro spinning process in the near future. This process has to ensure that the generated silk fiber resembles natural silk in its microstructure, chemical composition and mechanical properties (Fig. 4).[46]

To adapt the sophisticated spinning machinery of spiders, several aspects have to be taken into account. Importantly, besides the protein composition of the spinning dope and the phase separation process in the spinning duct, several mechanical parameters play critical roles in silk assembly. In nature, spiders draw the thread with the hind legs (or by the force of gravity in case of roping) out of the spinning wart.[15,36,37] This drawing process has been copied in the laboratory by forced silking of captive spiders. Analysis of the silking process revealed that spinning speeds range from 0.1 to 400 mm per second. Interestingly, large differences in resilience, ductility and thread diameters were reported depending on spinning speed and temperature.[55] It was shown that silks produced at higher reeling speeds have a slightly higher yield point, but are less extensible and weaker than silks spun at lower speeds.

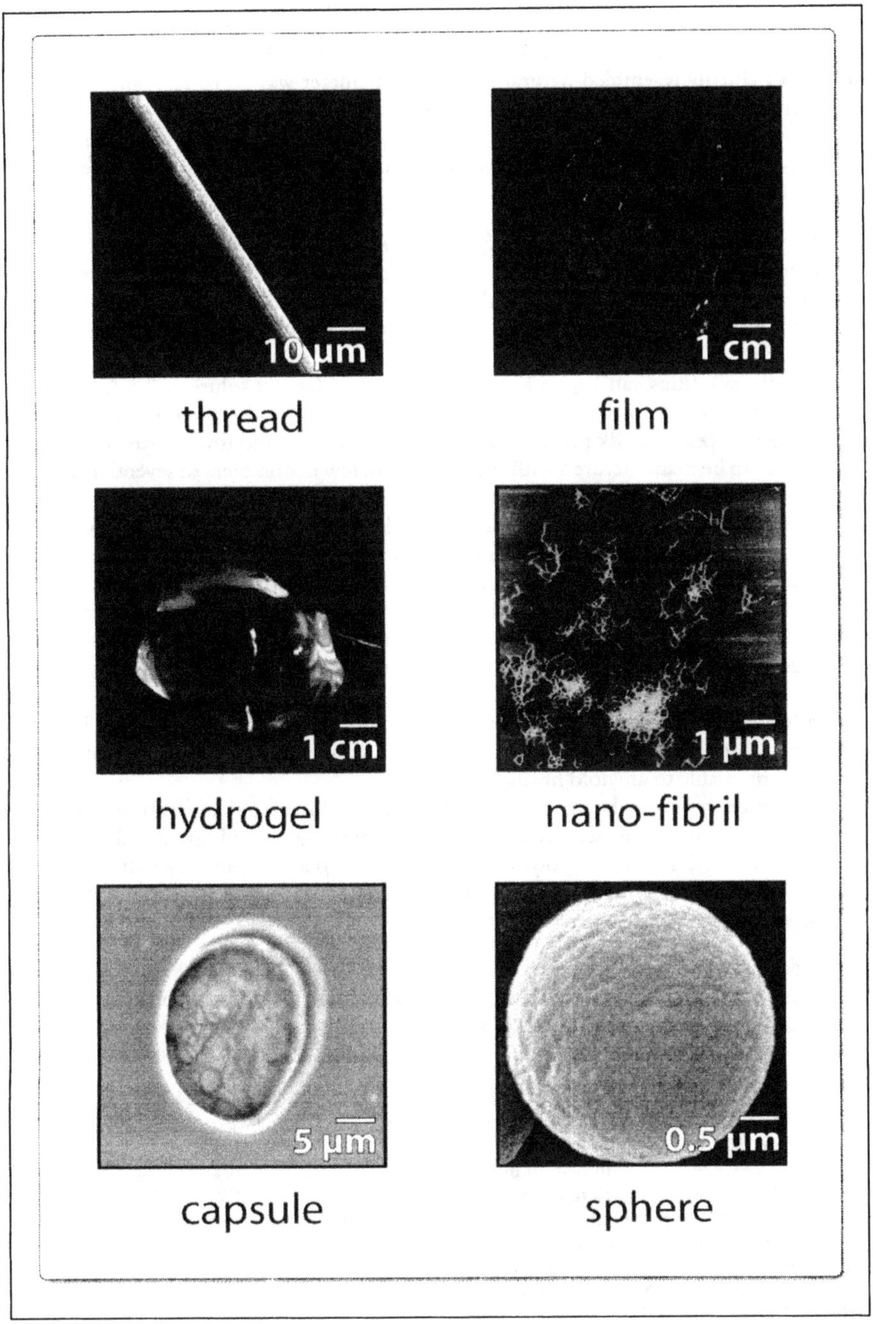

Figure 5. Possible shapes of spider silk. In nature, spider silk proteins are exclusively converted into silk threads. However, in vitro it is possible to transform silk proteins in other two- or three- dimensional shapes. The figure shows images of a capsule, a sphere, a thread and nano-fibrils (all electron microscopy), as well as a hydrogel and a film (photograph) made by recombinantly produced engineered spider silk protein.[35,42,48,71]

Several attempts have been made to spin recombinant spider silk proteins for scientific purposes. At first wet-spinning processes were employed with promising results.[67,68] Using silicon micro-spinnerets several meters of insect or spider silk fibers could be produced. Although the wet-spun silks visually resembled natural ones, their diameter was up to ten times bigger than naturally occurring silks leading to worse mechanical properties than found in natural silk fibers. Other attempts using special postspinning techniques yielded silks with even larger diameters.[62,69] Until now, even the best mechanical properties obtained by artificial spinning techniques are much lower than that of natural dragline silks.[68]

Exceeding Nature—Proteins as Polymer Materials

Spiders mainly employ silk as linear thread. However, spider silk proteins have much more potential, because they can assemble in all three dimensions to build distinct macroscopic structures (Fig. 5). Nowadays, several attempts are underway to employ silk proteins as biopolymer for novel materials.

For example, silk films can be made from a diluted spider silk solution.[70,71] A silk protein solution is poured onto a surface and the solvent (e.g., water) is allowed to evaporate. During the evaporation process, the silk proteins assemble on the surface and form a sturdy, transparent film. The films can be manufactured with thickness from few nanometers to several micrometers sharing different chemical and mechanical characteristics, depending on the choice of solvent and environmental conditions. Although silk films themselves cannot fulfill the role of model systems for the assembly behavior of all spider silk proteins, they give valuable insights into secondary and tertiary structure formation of these proteins under laboratory conditions. Analysis confirmed that MA proteins are intrinsically unfolded while in aqueous solution. Upon film formation, the proteins rapidly switch to a helical structure. Post treatment of the films with organic solvents like methanol resulted in further rearrangements of protein structure, increasing the β-sheet content dramatically.[71-73]

In vitro spider silk is further able to self-assemble into small nano-fibrils upon incubation in potassium phosphate buffer for several days at room temperature.[74,75] The resulting fibrils are structurally comparable to amyloid fibrils.

Taken together, analyzing the structure-function relationship of spider silk proteins in the near future will not only unravel the "secrets" behind the extreme toughness of silk threads, but will also help to engineer and design novel polymeric materials which are inspired by nature. The direct control of silk assembly will enable scientists to manufacture tailor-made spider silk materials with outstanding properties whose diversity cannot be achieved by conventional materials.

Acknowledgement

We thank John Hardy and Eileen Lintz for critical comments on the manuscript. This work was supported by the DFG (SCHE 603/4-3).

References

1. Gerritsen VB. An airbus could tiptoe on spider silk. Protein Spotlight 2000; 24:1-2.
2. Kaplan D, Adams WW, Farmer B et al. Silk polymers: material science and biotechnology. 1st ed. Washington DC: ACS Symposium Series, 1993.
3. Fraser RD, MacRae TP. Conformation in Fibrous Proteins. 1 ed. New York: Academic Press, 1973.
4. Vollrath F. Strength and structure of spiders' silks. J Biotechnol 2000; 74(2):67-83.
5. Nentwig W. Why do only certain insects escape from a spider's web? Oecologica 1982; 53:412-417.
6. Zschokke S. The influence of the auxiliary spiral on the capture spiral in Araneus diadematus Clerck (Araneidae). Bull Br Arachnol Soc 1993; 9(6):167-173.
7. Gosline JM, DeMont EM, Denny MW. The structure and properties of spider silk. Endeavour 1986; 10:37-43.
8. Hu X, Yuan J, Wang X et al. Analysis of aqueous glue coating proteins on the silk fibers of the cob weaver, Latrodectus hesperus. Biochemistry 2007; 46(11):3294-3303.
9. Gao H, Yao H. Shape insensitive optimal adhesion of nanoscale fibrillar structures. Proc Natl Acad Sci USA 2004; 101(21):7851-7856.

10. Arzt E, Gorb S, Spolenak R. From micro to nano contacts in biological attachment devices. Proc Natl Acad Sci USA 2003; 100(19):10603-10606.
11. Townley MA, Tillinghast EK, Neefus CD. Changes in composition of spider orb web sticky droplets with starvation and web removal and synthesis of sticky droplet compounds. J Exp Biol 2006; 209(Pt 8):1463-1486.
12. Vollrath F, Tillinghast EK. Glycoprotein glue beneath a spider web's aqueous coat. Naturwissenschaften 2005; 78(12):557-559.
13. Rising A, Nimmervoll H, Grip S et al. Spider silk proteins—mechanical property and gene sequence. Zoolog Sci 2005; 22(3):273-281.
14. Xu M, Lewis RV. Structure of a protein superfiber: spider dragline silk. Proc Natl Acad Sci USA 1990; 87(18):7120-7124.
15. Lewis RV. Spider silk: ancient ideas for new biomaterials. Chem Rev 2006; 106(9):3762-3774.
16. Hayashi CY, Lewis RV. Evidence from flagelliform silk cDNA for the structural basis of elasticity and modular nature of spider silks. J Mol Biol 1998; 275(5):773-784.
17. Hayashi CY, Lewis RV. Spider flagelliform silk: lessons in protein design, gene structure and molecular evolution. Bioessays 2001; 23(8):750-756.
18. van Beek JD, Hess S, Vollrath F et al. The molecular structure of spider dragline silk: folding and orientation of the protein backbone. Proc Natl Acad Sci USA 2002; 99(16):10266-10271.
19. Simmons AH, Michal CA, Jelinski LW. Molecular orientation and two-component nature of the crystalline fraction of spider dragline silk. Science 1996; 271(5245):84-87.
20. Gosline JM, Denny MW, DeMont EM. Spider silk as rubber. Nature 1994; 309:551-552.
21. Termonia Y. Monte Carlo diffusion model of polymer coagulation. Phys Rev Lett 1994; 72(23):3678-3681.
22. Becker N, Oroudjev E, Mutz S et al. Molecular nanosprings in spider capture-silk threads. Nat Mater 2003; 2(4):278-283.
23. Scheibel T. Protein fibers as performance proteins: new technologies and applications. Curr Opin Biotechnol 2005; 16(4):427-433.
24. Rising A, Hjalm G, Engstrom W et al. N-terminal nonrepetitive domain common to dragline, flagelliform and cylindriform spider silk proteins. Biomacromolecules 2006; 7(11):3120-3124.
25. Ittah S, Michaeli A, Goldblum A et al. A model for the structure of the C-terminal domain of dragline spider silk and the role of its conserved cysteine. Biomacromolecules 2007; 8(9):2768-2773.
26. Sponner A, Vater W, Rommerskirch W et al. The conserved C-termini contribute to the properties of spider silk fibroins. Biochem Biophys Res Commun 2005; 338(2):897-902.
27. Sponner A, Unger E, Grosse F et al. Conserved C-termini of Spidroins are secreted by the major ampullate glands and retained in the silk thread. Biomacromolecules 2004; 5(3):840-845.
28. Huemmerich D, Helsen CW, Quedzuweit S et al. Primary structure elements of spider dragline silks and their contribution to protein solubility. Biochemistry 2004; 43(42):13604-13612.
29. Hu X, Vasanthavada K, Kohler K et al. Molecular mechanisms of spider silk. Cell Mol Life Sci 2006; 63(17):1986-1999.
30. Motriuk-Smith D, Smith A, Hayashi CY et al. Analysis of the conserved N-terminal domains in major ampullate spider silk proteins. Biomacromolecules 2005; 6(6):3152-3159.
31. Ayoub NA, Garb JE, Tinghitella RM et al. Blueprint for a high-performance biomaterial: full-length spider dragline silk genes. PLoS ONE 2007; 2:e514.
32. Castillo-Davis CI, Mekhedov SL, Hartl DL et al. Selection for short introns in highly expressed genes. Nat Genet 2002; 31(4):415-418.
33. Exler JH, Hummerich D, Scheibel T. The amphiphilic properties of spider silks are important for spinning. Angew Chem Int Ed Engl 2007; 46(19):3559-3562.
34. Hermanson KD, Huemmerich D, Scheibel T et al. Engineered Microcapsules Fabricated from Reconstituted Spider Silk. Advanced Materials 2007; 19(14):1810-1815.
35. Jin HJ, Kaplan DL. Mechanism of silk processing in insects and spiders. Nature 2003; 424(6952):1057-1061.
36. Vollrath F, Knight DP. Liquid crystalline spinning of spider silk. Nature 2001; 410(6828):541-548.
37. Riekel C, Branden C, Craig C et al. Aspects of X-ray diffraction on single spider fibers. Int J Biol Macromol 1999; 24(2-3):179-186.
38. Gosline JM, Guerette PA, Ortlepp CS et al. The mechanical design of spider silks: from fibroin sequence to mechanical function. J Exp Biol 1999; 202(Pt 23):3295-3303.
39. Kubik S. High-performance fibers from spider silk. Angew Chem Int Ed Engl 2002; 41(15):2721-2723.
41. Hayashi CY, Shipley NH, Lewis RV. Hypotheses that correlate the sequence, structure and mechanical properties of spider silk proteins. Int J Biol Macromol 1999; 24(2-3):271-275.

42. Rammensee S, Huemmerich D, Hermanson KD et al. Rheological characterisation of recombinant spider silk nanofiber networks. Appl Phys A 2006; 82:261-264.
43. Dicko C, Kenney JM, Knight D et al. Transition to a beta-sheet-rich structure in spidroin in vitro: the effects of pH and cations. Biochemistry 2004; 43(44):14080-14087.
44. Hijirida DH, Do KG, Michal C et al. 13C NMR of Nephila clavipes major ampullate silk gland. Biophys J 1996; 71(6):3442-3447.
45. SenGupta S, Scheibel T. Folding, self-assembly and conformational switches of proteins. In: Zbilut JP, Scheibel T, eds. Protein folding and misfolding. New York: Nova Publishers, 2007:1-34.
46. Scheibel T. Spider silks: recombinant synthesis, assembly, spinning and engineering of synthetic proteins. Microb Cell Fact 2004; 3(1):14.
47. Ko FK, Jovicic J. Modeling of mechanical properties and structural design of spider web. Biomacromolecules 2004; 5(3):780-785.
48. Vendrely C, Scheibel T. Biotechnological production of spider-silk proteins enables new applications. Macromol Biosci 2007; 7(4):401-409.
49. Emile O, Le Floch A, Vollrath F. Biopolymers: shape memory in spider draglines. Nature 2006; 440(7084):621.
50. Emile O, Le Floch A, Vollrath F. Time-resolved torsional relaxation of spider draglines by an optical technique. Phys Rev Lett 2007; 98(16):167402.
51. Liu Y, Shao Z, Vollrath F. Relationships between supercontraction and mechanical properties of spider silk. Nat Mater 2005; 4(12):901-905.
52. Perez-Rigueiro J, Elices M, Guinea GV. Controled supercontraction tailors the tensile behaviour of spider silk. Polymer 2003; 44:3733-3736.
53. Shao Z, Vollrath F, Sirichaisit J et al. Analysis of spider silk in native and supercontracted states using raman spectroscopy. Polymer 1999; 40:2493-2500.
54. Yang Z. Supercontraction and backbone dynamics in spider silk: C-12 and H-2 NMR studies. J Am Chem Soc 2000; 122:9019-9025.
55. Vollrath F, Madsen B, Shao Z. The effect of spinning conditions on the mechanics of a spider's dragline silk. Proc Biol Sci 2001; 268(1483):2339-2346.
56. Wilding MA, Hearle J. Fiber structure. In: Salamone JC, eds. Polymeric materials encyclopedia vol. 11. Bota Raton: CRC, 1996: 8307-8322.
57. Shao Z, Vollrath F. Surprising strength of silkworm silk. Nature 2002; 418(6899):741.
58. Arcidiacono S, Mello C, Kaplan D et al. Purification and characterization of recombinant spider silk expressed in Escherichia coli. Appl Microbiol Biotechnol 1998; 49(1):31-38.
59. Fahnestock SR, Bedzyk LA. Production of synthetic spider dragline silk protein in Pichia pastoris. Appl Microbiol Biotechnol 1997; 47(1):33-39.
60. Scheller J, Guhrs KH, Grosse F et al. Production of spider silk proteins in tobacco and potato. Nat Biotechnol 2001; 19(6):573-577.
61. Trivedi BP, Lab spins artificial spider silk, paving the way to new materials, http://news.nationalgeographic.com/news/2002/01/0117_020117TVspidermammals.html, 2002.
62. Lazaris A, Arcidiacono S, Huang Y et al. Spider silk fibers spun from soluble recombinant silk produced in mammalian cells. Science 2002; 295(5554):472-476.
63. Miao Y, Zhang Y, Nakagaki K et al. Expression of spider flagelliform silk protein in Bombyx mori cell line by a novel Bac-to-Bac/BmNPV baculovirus expression system. Appl Microbiol Biotechnol 2006; 71(2):192-199.
64. Huemmerich D, Scheibel T, Vollrath F et al. Novel assembly properties of recombinant spider dragline silk proteins. Curr Biol 2004; 14(22):2070-2074.
65. Padgett KA, Sorge JA. Creating seamless junctions independent of restriction sites in PCR cloning. Gene 1996; 168(1):31-35.
66. Schmidt M, Romer L, Strehle M et al. Conquering isoleucine auxotrophy of Escherichia coli BLR(DE3) to recombinantly produce spider silk proteins in minimal media. Biotechnol Lett 2007; 29(44):1741-1744.
67. Jelinski LW, Blye A, Liivak O et al. Orientation, structure, wet-spinning and molecular basis for supercontraction of spider dragline silk. Int J Biol Macromol 1999; 24(2-3):197-201.
68. Seidel A, Liivak O, Jelinski LW. Artificial spinning of spider silk. Macromolecules 1998; 31:6733-6736.
69. Arcidiacono S, Mello CM, Butler M et al. Aqueous processing and fiber spinning of recombinant silks. Macromolecules 2002; 35:1262-1266.
70. Ackerschott C, Vendrely C, Romer L et al. Modular design of recombinant repetitive proteins: flagelliform spider silk as basis for biomaterials. Protocols in Nanostructural Design 2008, in press.
71. Slotta U, Tammer M, Kremer F et al. Structural analysis of spider silk films. Supramol Chem 2006; 18(5):465-471.

72. Metwalli E, Slotta U, Darko C et al. Structural changes of thin films from recombinant spider silk proteins upon post treatment. Appl Phys A 2007; 89:655-661.
73. Huemmerich D, Slotta U, Scheibel T. Films from recombinant spider silk proteins. Appl Phys A 2006; 82:219-222.
74. Kenney JM, Knight D, Wise MJ et al. Amyloidogenic nature of spider silk. Eur J Biochem 2002; 269(16):4159-4163.
75. Slotta U, Hess S, Spiess K et al. Spider silk and amyloid fibrils: a structural comparison. Macromol Biosci 2007; 7(2):183-188.

Index

For Product Safety Concerns and Information please contact our EU representative GPSR@taylorandfrancis.com
Taylor & Francis Verlag GmbH, Kaufingerstraße 24, 80331 München, Germany

www.ingramcontent.com/pod-product-compliance
Lightning Source LLC
LaVergne TN
LVHW021339191026
850543LV00004B/626
* 9 7 8 1 5 8 7 0 6 3 1 6 9 *